TENSORFLOW
Pocket Primer

TENSORFLOW
Pocket Primer

Oswald Campesato

MERCURY LEARNING AND INFORMATION

Dulles, Virginia
Boston, Massachusetts
New Delhi

Publisher: David Pallai
Mercury Learning and Information
22841 Quicksilver Drive
Dulles, VA 20166
info@merclearning.com
www.merclearning.com
800-232-0223

O. Campesato. *TensorFlow Pocket Primer.*
ISBN: 978-1-68392-364-0

The publisher recognizes and respects all marks used by companies, manufacturers, and developers as a means to distinguish their products. All brand names and product names mentioned in this book are trademarks or service marks of their respective companies. Any omission or misuse (of any kind) of service marks or trademarks, etc. is not an attempt to infringe on the property of others.

Library of Congress Control Number: 2019939376

192021321   This book is printed on acid-free paper in the United States of America.

Our titles are available for adoption, license, or bulk purchase by institutions, corporations, etc. For additional information, please contact the Customer Service Dept. at 800-232-0223 (toll free).

*I'd like to dedicate this book to my parents —
may this bring joy and happiness into their lives.*

CONTENTS

PREFACE

WHAT IS THE GOAL?

The goal of this book is to introduce advanced beginners to TensorFlow 1.x fundamentals for basic machine learning algorithms in TensorFlow. It is intended to be a fast-paced introduction to various "core" features of Tensor-Flow, with code samples that cover deep learning and TensorFlow. The material in the chapters illustrates how to solve a variety of tasks using TensorFlow, after which you can do further reading to deepen your knowledge.

This book provides more detailed code samples than those that are found in intermediate and advanced TensorFlow books. Although it contains some basic code samples in TensorFlow, some familiarity with TensorFlow will be helpful.

The book will also save you the time required to search for code samples, which is a potentially time-consuming process. In any case, if you're not sure whether or not you can absorb the material in this book, glance through the code samples to get a feel for the level of complexity.

At the risk of stating the obvious, please keep in mind the following point: *you will not become an expert in TensorFlow by reading this book.*

WHAT WILL I LEARN FROM THIS BOOK?

The first chapter contains TensorFlow code samples that illustrate very simple TensorFlow functionality, followed by a chapter where code samples illustrate an assortment of TensorFlow built-in APIs. The third chapter delves into the TensorFlow Dataset, with a plethora of code samples that illustrate how to use "lazy" operators in conjunction with datasets. The fourth chapter discusses linear regression and the fifth chapter delves into logistic regression. If you think that you'll struggle significantly with the code in the first two chapters, then an "absolute beginners" type of book is recommended to prepare you for this book.

Another point: although Jupyter is popular, all the code samples in this book are Python scripts. However, you can quickly learn about the useful features of Jupyter through various online tutorials. In addition, it's worth looking at Google Collaboratory, which is entirely online and is based on Jupyter notebooks, along with free GPU usage.

HOW WERE THE CODE SAMPLES CREATED?

The code samples in this book were created and tested using TensorFlow version 1.12 on a Macbook Pro with OS X 10.12.6 (macOS Sierra). Regarding their content: the code samples are derived primarily from the author for his "Deep Learning and TensorFlow" graduate course. In some cases, there are code samples that incorporate short sections of code from discussions in online forums. The key point to remember is that the code samples follow the "Four Cs": they must be clear, concise, complete, and correct to the extent that it's possible to do so, given the size of this book.

WHAT ARE THE TECHNICAL PREREQUISITES FOR THIS BOOK?

You need some familiarity with Python, and also know how to launch Python code from the command line (in a Unix-like environment for Mac users). In addition, a mixture of basic linear algebra (vectors and matrices), probability/ statistics (mean, median, standard deviation), and basic concepts in calculus (such as derivatives) will help you learn the material in this book.

The following prerequisite is important for understanding the code samples in the last two chapters of this book: some familiarity with neural networks, which includes the concept of hidden layers and activation functions (even if you don't fully understand them). Knowledge of cross entropy is also helpful for some of the code samples. To phrase things more colloquially, this book assumes that you know how to crawl in order to help you learn how to walk.

Also keep in mind that TensorFlow provides a vast assortment of APIs, some of which are discussed in the code samples in the book chapters. While it's possible for you to "pick up" the purpose of those APIs by reading the online documentation, that's only true for the basic TensorFlow APIs, which is to say that you probably won't really understand how they work and why they are needed until you work with them in TensorFlow code samples. *In other words, if you read TensorFlow code samples containing APIs that you do not understand, in many cases you will not understand those APIs even after repeatedly reading the code samples.*

A more efficient approach is to learn about the purpose of the TensorFlow APIs by reading small code samples that illustrate the purpose of those APIs, after which you can read more complex TensorFlow code samples.

WHAT ARE THE NON-TECHNICAL PREREQUISITES FOR THIS BOOK?

Although the answer to this question is more difficult to quantify, it's very important to have a strong desire to learn TensorFlow and machine learning, along with the motivation and discipline to read and understand the code samples.

Even the non-trivial TensorFlow APIs can be a challenge to understand the first time you encounter them, so be prepared to read the code samples several times. The latter requires persistence when learning TensorFlow, and whether or not you have enough persistence is something that you need to decide for yourself.

WHICH TOPICS ARE EXCLUDED?

This book does not cover Convolutional Neural Networks (CNN), Recurrent Neural Networks (RNN), or Long Short Term Memory (LSTM). You will not find in-depth details about TensorFlow layers and estimators (but they are lightly discussed). Keep in mind that online searches on Stackoverflow will often contain solutions involving different versions of TF 1.x.

HOW DO I SET UP A COMMAND SHELL?

If you are a Mac user, there are three ways to do so. The first method is to use Finder to navigate to Applications > Utilities and then double click on the Utilities application. Next, if you already have a command shell available, you can launch a new command shell by typing the following command:

```
open /Applications/Utilities/Terminal.app
```

A second method for Mac users is to open a new command shell on a Macbook from a command shell that is already visible simply by clicking command+n in that command shell Your Mac will then launch another command shell.

If you are a PC user, you can install Cygwin (open source *https://cygwin.com/*) that simulates bash commands, or use another toolkit, such as MKS (a commercial product). Please read the online documentation that describes the download and installation process. Note that custom aliases are not automatically set if they are defined in a file other than the main start-up file (such as .bash_login).

WHAT ARE THE "NEXT STEPS" AFTER FINISHING THIS BOOK?

The answer to this question varies widely, mainly because the answer depends heavily on your objectives. The best answer is to try a new tool or technique from the book on a problem or task you care about, professionally or personally. Precisely what that might be depends on who you are, as the needs of a data scientist, manager, student, or developer are all different. In addition, keep what you learned in mind as you tackle new challenges.

If you have reached the limits of what you have learned here and want to get further technical depth regarding TensorFlow, there are various online resources and literature describing more complex features of TensorFlow.

Unless TensorFlow 1.x is a requirement, it's probably a good idea to set aside some time to learn TensorFlow 2, which is available from the author in an upcoming Pocket Primer. Incidentally, another option to consider is Keras that is located in `tf.keras`, which provides a layer of abstraction on top of TensorFlow that will enable you to develop prototypes more quickly than in TensorFlow.

COMPANION FILES

Companion files with code samples and images from the book may be obtained by writing to the publisher at info@merclearning.com.

## CHAPTER 1

INTRODUCTION TO TENSORFLOW

This chapter provides a quick introduction to various features of Tensor-Flow, and some of the TensorFlow tools and projects that are included in the TensorFlow "family." The intent of this chapter is to provide relevant TensorFlow information, how to visualize TensorFlow graphs via Tensor-Board, how to invoke TensorFlow code in a browser via Jupyter notebooks, and where to train neural networks with free GPU support (Google Colaboratory). The material in this chapter will prepare you for Chapter 2, which provides a "foundation" of commonly used TensorFlow APIs (illustrated via short code samples) that are also in code samples in the remaining chapters of this book.

There are a few points to keep in mind before you read this chapter. First, in this book, TensorFlow refers to TensorFlow 1.x, unless explicitly stated otherwise. For expediency, you will often see the acronym "TF" used instead of TensorFlow. In addition, the historical details regarding TensorFlow are minimized in order to provide you with a decent set of TensorFlow code samples.

Second, the code samples in this book were tested on a Macbook Pro (12.5.3) with Python 3.5.1 and TensorFlow 1.12, which is currently the latest version of TensorFlow 1.x. When TensorFlow 2 is released in 2019, all TensorFlow 1.x releases will become legacy code. However, Google will probably continue supporting TensorFlow 1.x for at least another year after the release of TensorFlow 2. At that point there won't be any new code development for TF 1.x (except for security-related updates).

The first part of this chapter briefly discusses some TensorFlow features and some of the tools that are part of the TensorFlow "family." The second section of this chapter shows you how to write TensorFlow code that contains various combinations of TF constants, TF placeholders, and TF variables. After reading these code samples, you will learn about the differences between TF placeholders and TF variables, which tend to confuse people who are new to TensorFlow.

The third section of this chapter shows you how to perform arithmetic operations in TensorFlow, how to use various built-in functions, how to calculate trigonometric values, `for` loops, `while` loops, the `tf.less()` API, and how to calculate exponential values.

The fourth section contains TF code samples that perform various operations on arrays, such as creating an identity matrix, a constant matrix, a random uniform matrix, and a truncated normal matrix. This section also shows you how to multiply TensorFlow arrays and how to convert Python arrays to TensorFlow arrays.

The fourth section shows you how to save a TensorFlow graph so that you can view its contents in a browser using the built-in TensorBoard utility that is launched from the command line. The final section introduces you to Google Colaboratory, which is a fully online Jupyter-based environment that offers 12 hours of daily GPU usage for free.

WHAT IS TENSORFLOW?

TensorFlow is an open source framework from Google that was released in November, 2015. The TensorFlow framework is for machine learning and deep learning. TensorFlow evolved from Google Brain and is available through an Apache license. In this book, the TensorFlow code samples use Python 3.x. TensorFlow also supports a variety of programming languages and hardware platforms. Here is a short list of some TensorFlow features:

Support for Python, Java, C++
Desktop, server, mobile device (TF Lite)
CPU/GPU/TPU support
Linux and Mac OS X support
VM for Windows

Navigate to the TensorFlow home page, where you will find links to many resources for TensorFlow:
https://www.tensorflow.org
Install TensorFlow by issuing the following command from the command line:

```
pip install tensorflow
```

If you want to upgrade TensorFlow to the latest version, issue the following command from the command line:

```
pip install --upgrade tensorflow
```

TensorFlow Architecture (High View)

TensorFlow consists of two main components: 1) a graph protocol buffer, and 2) a runtime to execute (distributed) graph that is analogous to Python code and the Python interpreter. TensorFlow is written in C++ and supports various graph-based operations involving primitive values and so-called tensors (discussed later).

TensorFlow creates a "computation" graph for numerical computation and data flow, with the notion that "everything is a graph." Graph and data visualization are handled via TensorBoard (discussed later) that is included as part of TensorFlow. As you will see in the code samples in this book, TensorFlow APIs are available in Python and can therefore be embedded in Python scripts.

The default execution mode for TF 1.x is *deferred execution*, whereas TF 2 uses *eager execution* (think "immediate mode"). Although TF 1.4 introduced eager execution, the vast majority of TF 1.x code samples that you will find online uses deferred execution. TensorFlow supports arithmetic operations on tensors (i.e., multi-dimensional arrays with enhancements) as well as conditional logic and `while` loops.

TensorFlow provides a "checkpoint" API to save TensorFlow graphs and restore them at a later point. In addition, you can view saved graphs in TensorBoard, which is a very useful data visualization tool that is bundled with TensorFlow.

TensorFlow Features

The following list contains various high-level features of TensorFlow:

Distributed computation
R/D for developing new ML algorithms
A REPL environment
Take models from training to production
Large-scale distributed models
TF graph computed on different machines
Models for mobile

Most of the preceding items are self-explanatory. The TensorFlow REPL (read-eval-print-loop) is available through the Python REPL, which is accessible by opening a command shell and then typing the following command:

```
python
```

As a simple illustration of accessing TensorFlow-related functionality in the Python REPL, import the TensorFlow library as follows:

```
>>> import tensorflow as tf
```

Now check the version of TensorFlow that is installed on your machine with this command:

```
>>> print('TF version:',tf.__version__)
```

The output of the preceding code snippet is shown here (the number that you see depends on which version of TensorFlow you installed):

```
TF version: 1.12.0
```

Although the Python REPL is useful for short code blocks, it's simpler to place the TensorFlow code samples in this book inside Python scripts that you can launch with the Python executable.

TensorFlow Use Cases

TensorFlow is designed to solve many use cases, some of which are listed here:

Image recognition
Computer vision
Voice/sound recognition
Time series analysis
Language detection
Language translation
Text-based processing
Handwriting recognition

In case you didn't already know, the preceding list also includes common-use cases that are suitable for machine learning as well as deep learning.

OTHER TENSORFLOW-BASED TOOLKITS

TensorFlow has the following associated toolkits:

TensorBoard (included as part of TensorFlow)
TensorFlow Serving (hosting)
TensorFlow Lite (for mobile)
tensorflow.js (for web pages and NodeJS)

The TensorFlow distribution contains TensorBoard, which is a graph visualization tool that runs in a browser. For example, type the following command that accesses a saved TF graph in the subdirectory /tmp/abc:

```
tensorboard -logdir=/tmp/abc
```

Next, open a browser session and navigate to this URL:

```
localhost:6006
```

Now you will see a visualization of the TensorFlow graph that was saved as a file in the directory /tmp/abc.

TensorFlow Serving is a flexible, high-performance serving system for machine learning models that is designed for production environments. TensorFlow Serving makes it easy to deploy new algorithms and experiments, while keeping the same server architecture and APIs. More information is here:

https://www.tensorflow.org/serving/

TensorFlow Lite was created for developing mobile applications (both Android and iOS). TensorFlow Lite supersedes TensorFlow Mobile, which was an earlier SDK for developing mobile applications. TensorFlow Lite is a lightweight solution for mobile and embedded devices. It enables on-device machine learning inference with low latency and a small binary size. TensorFlow Lite also supports hardware acceleration with the Android Neural Networks API. More information about TensorFlow Lite is here:

https://www.tensorflow.org/lite/

A more recent addition is `tensorflow.js` that provides JavaScript APIs to access TensorFlow in a Web page. The `tensorflow.js` toolkit was previously called `deeplearning.js`. In addition, tensorflow.js can be used with NodeJS. More information about `tensorflow.js` is here:

https://js.tensorflow.org/

WHAT ABOUT TENSORFLOW 2?

TensorFlow 2 will probably be released around mid-2019, at which point TensorFlow 1.x will become legacy code (as mentioned in the introduction). One of the most significant changes involves "eager" execution as the default mode instead of "deferred" execution mode.

Why use TF 1.x Instead of TF 2?

There are several reasons for staying with TF 1.x instead of switching to TF 2. First, as this book goes to print, TF 2 has not been released. Second (and perhaps even more important), currently there is a very small percentage of online TensorFlow code samples that use eager execution. Fortunately, Google will provide tools for converting TF 1.x code to TF 2 that will assist in many situations, i.e., when there is a straightforward conversion from TF 1.x APIs to TF 2 APIs. When those tools cannot perform the conversion, there will probably be an assortment of online tutorials available that will explain how to make the conversion.

Third, there is a massive code base that uses TF 1.x, which includes many customers and even Google itself. The conversion process from TF 1.x to TF 2 will take a lot of time because some (probably large) companies will be slow to make the transition. As an analogy, consider the migration from Python 2.x to Python 3.x: there are still many places that continue using Python 2.x, even though Python 3.x was released several years ago (and also the "end of life" for Python 2.x in 2020).

There is one possible exception: if you are "brand new" to TensorFlow and you do not have an urgent need to learn TensorFlow, then it might be less work for you to wait until a production version of TF 2 is available.

WHAT IS A TENSORFLOW TENSOR?

In simplified terms, a TF tensor is an n-dimensional array that is similar to a NumPy ndarray. A TF tensor is defined by its dimensionality, as summarized here:

```
scalar number:        a zeroth-order tensor
vector:               a first-order tensor
matrix:               a second-order tensor
3-dimensional array:  a 3rd order tensor
```

As you will see in subsequent sections, TensorFlow supports various data types and primitive types. In addition to constants and variables, TensorFlow provides a data type called a "placeholder," which acts as a buffer that holds transient data. Use TF variables for things that need to be trained, such as the slope m and intercept b of a best-fitting line in the plane.

TensorFlow Data Types

TensorFlow supports the following data types:

- tf.float32
- tf.float64
- tf.int8
- tf.int16
- tf.int32
- tf.int64
- tf.uint8
- tf.string
- tf.bool

The data types in the preceding list are self-explanatory: two floating point types, four integer types, one unsigned integer type, one string type, and one Boolean type. As you can see, there is a 32-bit and a 64-bit floating point type, and integer types that range from 8-bit through 64-bit.

TensorFlow Primitive Types

TensorFlow supports the following primitive types (arrays are discussed later):

TF Constants
TF Placeholders
TF Variables

A TensorFlow *constant* is an immutable value, and a simple example is shown here:

```
aconst = tf.constant(3.0)
```

A TensorFlow *placeholder* allocates space for data, and it's somewhat analogous to a "buffer" in other languages, such as C. The really nice aspect of TF placeholders is that you do *not* need to specify their shape: they "automagically" assume the shape of the data that is assigned to them. A simple example of a TensorFlow placeholder is shown here:

```
a = tf.placeholder("float")
b = tf.placeholder("float")
c = tf.multiply(a,b)
```

As you can see in the preceding code block, c is defined as the product of a and b, which do not have a value. Use a `feed_dict` to assign values to placeholder, as shown here:

```
a = tf.placeholder("float")
b = tf.placeholder("float")
c = tf.multiply(a,b)

# assign values to a and b:
feed_dict = {a:2, b:3}
```

The calculation of c itself is shown later in the section that discusses TF sessions.

A TensorFlow *variable* is a "trainable value" in a TensorFlow graph. For example, the slope m and y-intercept b of a best-fitting line in the plane are two examples of trainable values. Some examples of TF variables are shown here:

```
b = tf.Variable(3, name="b")
x = tf.Variable(2, name="x")
z = tf.Variable(5*x, name="z")

W = tf.Variable(20)
lm = tf.Variable(W*x + b, name="lm")
```

Notice that b, x, and W are assigned numeric initial values, whereas z and W are defined as expressions. Specifically, the value of the variable z depends on the value of x (which equals 2), and the value of the variable lm depends on the values of W, x, and b. Both z and lm are evaluated inside a TensorFlow "session" block, as shown in a later section.

TENSORFLOW GRAPHS

Whenever you define TensorFlow code in a Python script, you will often use some combination of TF data types and other TF constructs. Before executing TensorFlow code, TensorFlow generates a graph structure that is based on the constants, placeholders, and variables that you have defined. A typical TensorFlow graph consists of the following:

Graph: graph of operations (DAG)
Sessions: contains graph(s)
lazy execution (default)
operations in parallel (default)
Nodes: operators/variables/constants
Edges: tensors

TF 1.x graphs are split into subgraphs and executed in parallel (or multiple CPUs or GPUs, if the latter are available).

The chapters in this book contain an assortment of TensorFlow code samples written in Python that illustrate how to perform arithmetic operations, calculate trigonometric values, and how to use eager execution.

The TensorFlow Version Number

Listing 1.1 displays the contents of tf-version.py that illustrates how to find the version number of TensorFlow that is installed on your machine.

LISTING 1.1: tf-version.py

```
import tensorflow as tf

print('TF version:',tf.__version__)
```

Listing 1.1 contains an import statement and one print() statement that displays the installed version of TensorFlow. The output from Listing 1.1 is here:

```
('TF version:', '1.12.0')
```

Listing 1.1 consists of two statements: an import statement for TensorFlow followed by a print() statement that displays the version of TensorFlow that is installed on your machine. The output from Listing 1.1 is here (the number depends on the version of TensorFlow that you installed on your machine):

```
TF version: 1.12.0
```

A TENSORFLOW GRAPH WITH TF.SESSION()

A Tensorflow graph in TF 1.x consists of two parts: TensorFlow code in the first portion of a Python script, followed by an instance of a tf.Session() object (often abbreviated as sess). In order to obtain values from TF tensors, you invoke the sess.run() method and specify a TF tensor as an argument to the `sess.run()` method. By contrast, TF 2 does not require an instance of `tf.Session()`, and it's more "Pythonesque" than TF 1.x.

In extremely simple cases of TF 1.x code, such as displaying the version number of TensorFlow, you do not need a `tf.Session()` object.

Listing 1.2 displays the contents of `tf-session.py` that illustrates how to use the `tf.Session()` object in a TF graph.

LISTING 1.2: tf-session.py

```
import tensorflow as tf

aconst = tf.constant(3.0)
print(aconst)
# output: Tensor("Const:0", shape=(), dtype=float32)

sess = tf.Session()
print(sess.run(aconst))
# output: 3.0

sess.close()
# => there's a better way...
```

Listing 1.2 starts with an `import` statement, the definition of the TF constant `aconst`, and one `print()` statement that displays the metadata for `aconst`. The next portion of Listing 1.2 initializes the variable sess as an instance of the `TF.Session` class, followed by a `print()` statement that display the value of the variable `aconst`. The output from Listing 1.2 is here:

```
3.0
```

Listing 1.3 displays the contents of `tf-const2.py` that illustrates how to use the `tf.Session()` object in a TF graph.

LISTING 1.3: tf-session2.py

```
import tensorflow as tf

aconst = tf.constant(3.0)
print(aconst)

# Automatically close "sess"
with tf.Session() as sess:
  print(sess.run(aconst))
```

Listing 1.3 contains almost the same code as Listing 1.2: the difference is the `with` code snippet, which does not require explicitly closing the TF session. The output from Listing 1.3 is here:

```
3.0
```

PLACEHOLDERS AND FEED_DICT IN A TF SESSION

Listing 1.4 displays the contents of `tf-ph-feeddict.py` that illustrates how to compute values involving TF placeholders with code block in TensorFlow.

LISTING 1.4: tf-ph-feeddict.py

```
import tensorflow as tf

a = tf.placeholder("float")
b = tf.placeholder("float")
c = tf.multiply(a,b)

# initialize a and b:
feed_dict = {a:2, b:3}

# multiply a and b:
with tf.Session() as sess:
  print(sess.run(c, feed_dict))
```

Listing 1.4 defines two TF placeholders a and b, followed by the TF variable c that is the product of a and b. The output of Listing 1.4 is 6, which is the product of the values assigned to the TF placeholders a and b.

There is another important detail about TF variables and how they are initialized in TensorFlow, which is the subject of the next section.

CONSTANTS AND VARIABLES IN A TF SESSION

Listing 1.5 displays the contents of `tf-variables-init.py` that illustrates how to compute values involving TF placeholders in a `with` code block in TensorFlow.

LISTING 1.5: tf-variables-init.py

```
import tensorflow as tf

x = tf.constant(5,name="x")
y = tf.constant(8,name="y")
z = tf.Variable(2*x+3*y, name="z")

init = tf.global_variables_initializer()

with tf.Session() as session:
  session.run(init)
  print 'z = ',session.run(z) # =>  z = 34
```

The two code snippets shown in bold in Listing 1.5 are required (you can replace `init` with a different string) whenever a `with` code block contains a

TF variable, such as the variable z in Listing 1.5. The output of Listing 1.5 is 34, which is the result of computing 2*x + 3*y.

This concludes the quick tour involving TensorFlow code that contains various combinations of TF constants, TF placeholders, and TF variables. The next few sections delve into more details regarding the TF primitive types that you saw in the preceding sections.

CONSTANTS IN TENSORFLOW (REVISITED)

Here is a short list of some properties of TensorFlow constants:

- initialized during definition
- cannot change its value ("immutable")
- can specify its name (optional)
- the type is required (ex: tf.float32)
- assigned value only once (can use feed_dict)
- are not modified during training

Some examples of TensorFlow constants:

```
d = tf.constant("3", name="a")
e = tf.constant([5,5,5], tf.float32)
```

Listing 1.6 displays the contents of `tf-constants.py` that illustrates how to assign and print the values of some TensorFlow constants.

LISTING 1.6: tf-constants.py

```
import tensorflow as tf

scalar = tf.constant(10)
vector = tf.constant([1,2,3,4,5])
matrix = tf.constant([[1,2,3],[4,5,6]])
cube   = tf.constant([[[1],[2],[3]],[[4],[5],[6]],
                                     [[7],[8],[9]]])

print(scalar.get_shape())
print(vector.get_shape())
print(matrix.get_shape())
print(cube.get_shape())
```

Listing 1.6 contains four `tf.constant()` statements that define TF tensors of dimension 0, 1, 2, and 3, respectively. The second part of Listing 1.6 contains four `print()` statements that display the shape of the four TF constants that are defined in the first section of Listing 1.6. The output from Listing 1.6 is here:

```
()
(5,)
(2, 3)
(3, 3, 1)
```

The tf.rank() API

The *rank* of a TensorFlow tensor is the dimensionality of the tensor, whereas the *shape* (discussed in the next section) of a tensor is the number of elements in each dimension. Listing 1.7 displays the contents of tf-rank.py that illustrates how to find the rank of TensorFlow tensors.

LISTING 1.7: tf-rank.py

```
import tensorflow as tf

# constant:
aconst = tf.constant(3.0)
print(aconst)

# 2x3 constant matrix
B = tf.fill([2,3], 5.0)

with tf.Session() as sess:
  print('aconst:',sess.run(tf.rank(aconst)))
  print('rank B:',sess.run(tf.rank(B)))
```

Listing 1.7 contains familiar code for defining the TF constant aconst, followed by the TF tensor B that is a 2x3 tensor (because of the tf.fill() API) in which every element of B has the value 5. The next block of code is a with code block that prints the values of aconst and the shape of B. The output from Listing 1.7 is here:

```
Tensor("Const:0", shape=(), dtype=float32)
aconst: 3.0
rank B: 2
```

The Shape of a TF Tensor

The *shape* of a TensorFlow tensor is the number of elements in each dimension of a given tensor. Listing 1.8 displays the contents of tf-shape.py that illustrates how to find the shape of TensorFlow tensors.

LISTING 1.8: tf-shape.py

```
import tensorflow as tf

# constant:
aconst = tf.constant(3.0)
print(aconst)

# 2x3 constant matrix
B = tf.fill([2,3], 5.0)

with tf.Session() as sess:
  print('aconst:',sess.run(tf.rank(aconst)))
  print('rank B:',sess.run(tf.rank(B)))
```

Listing 1.8 contains the usual `import` statement, followed by the definition of the TF constant `aconst`. Next, the TF variable B is initialized as a 1x2 vector that is initialized with the value 5.0. Next, a `with` code block contains the code to display the values of `aconst` and the rank of B. The output from Listing 1.8 is here:

```
Tensor("Const:0", shape=(), dtype=float32)
aconst: 3.0
rank B: 2
```

PLACEHOLDERS IN TENSORFLOW (REVISITED)

TensorFlow placeholders have the following features:

- they only allocate memory for future use
- they can have variable size
- used for inputting data "external" to graph
- good for unknown data size
- placeholders are from functions (not TF class instances)

The following examples of TensorFlow placeholders illustrate how to specify unconstrained shapes and higher-dimensional tensors:

```
a = tf.placeholder("float")
b = tf.placeholder(tf.int32, name='b')
c = tf.placeholder(tf.float32, shape=[3])

# Unconstrained shape:
w = tf.placeholder(tf.float32)

# Matrix of unconstrained size:
x = tf.placeholder(tf.float32, shape=[None, None])

# Matrix with 32 columns:
y = tf.placeholder(tf.float32, shape=[None, 32])

# 128x32-element matrix:
z = tf.placeholder(tf.float32, shape=[128, 32])
```

The following examples of TensorFlow placeholders illustrate how to specify constrained shapes and higher-dimensional tensors:

```
# placeholder with shape:
x = tf.placeholder(tf.float32,(3,4))

# feed any matrix with 4 columns and any number of rows at
#                                                run time:
x = tf.placeholder(tf.float32, shape=(None,4))

# placeholder X with unspecified number of rows of shape
#                          (128, 128, 3) of type float32:
X = tf.placeholder(tf.float32, shape=[None, 128, 128, 3],
                                                name="X")
```

TF PLACEHOLDERS AND FEED_DICT

Listing 1.9 displays the contents of `tf-feeddict-values.py` that illustrates how to specify values in a `feed_dict` in a TF graph.

LISTING 1.9: *tf-feeddict-values.py*

```
import tensorflow as tf

x = tf.placeholder("float", None)
y = x * 2

with tf.Session() as session:
  result = session.run(y, feed_dict={x: [1, 2, 3]})
  print('y:',result)
```

Listing 1.9 defines the TF placeholder x as a float data type, followed by the variable y that is twice the value of x. The `with` code block calculates the value of y by specifying an array of values for x in `feed_dict`. If you expected merely a scalar value for x, this example illustrates the fact that a TF placeholder can take whatever shape is required (i.e., not just a scalar value). The output from Listing 1.9 is here:

```
('y:', array([2., 4., 6.], dtype=float32))
```

Listing 1.10 displays the contents of `tf-feeddict-values2.py` that illustrates how to use `feed_dict` to supply values to a placeholder.

LISTING 1.10: *tf-feeddict-values2.py*

```
import tensorflow as tf

x = tf.placeholder("float", [None, 3])
y = x * 2
z = x ** 3

with tf.Session() as session:
  x_data = [[1, 2, 3],
            [4, 5, 6],]

  result1 = session.run(y, feed_dict={x: x_data})
  print('y:',result1.eval())

  result2 = session.run(z, feed_dict={x: x_data})
  print('z:',result2.eval())
```

Listing 1.10 defines the TF placeholder x as a float data type with an arbitrary number of rows and three columns. The variables y and z are defined as twice the value of x and x cubed, respectively.

The `with` code block defines the 2x3 array variable x_data with integer values, after which result1 is calculated by evaluating y with x_data for feed_dict. Similarly, result2 is calculated by evaluating z with x_data for feed_dict. The output from Listing 1.10 is here:

```
y: [[ 2.   4.   6.]
 [ 8. 10. 12.]]
z: [[  1.    8.   27.]
 [ 64. 125. 216.]]
```

VARIABLES IN TENSORFLOW (REVISITED)

In addition to constants and placeholders that are described earlier in this chapter, TensorFlow supports variables. Variables can be updated during backward error propagation (also called "backprop," which is discussed later). TF variables can also be saved in a graph and then restored at a later point in time.

TensorFlow also provides the method tf.assign() in order to modify values of TF variables. The following list contains some properties of TensorFlow variables:

• initial value is optional
• must be initialized before graph execution
• updated during training
• constantly recomputed
• they hold values for weights and biases
• in-memory buffer (saved/restored from disk)

Here are some examples of TensorFlow variables:

```
b = tf.Variable(3, name='b')
x = tf.Variable(2, name='x')
z = tf.Variable(5*x, name="z")

W = tf.Variable(20)
lm = tf.Variable(W*x + b, name="lm")
```

Notice that the variables b, x, and W specify constant values, whereas the variables z and lm specify expressions that are defined in terms of other variables. If you are familiar with linear regression, you undoubtedly noticed that the variable lm defines a line in the Euclidean plane.

Other properties of TensorFlow variables are listed below:

• a tensor that's updateable via operations
• exist outside the context of sess.run()
• like a "regular" variable
• holds the learned model parameters
• variables can be shared (or non-trainable)
• used for storing/maintaining state
• internally stores a persistent tensor

- modifications are visible across multiple `tf.Sessions`
- you can read/modify the values of the tensor
- multiple workers see the same values for `tf.Variables`
- the best way to represent shared, persistent state manipulated by your program

TF Variables versus TF Tensors

Keep in mind the following distinction between TF variables and TF tensors: TF *variables* represent your model's trainable parameters (ex: weights and biases of a neural network), whereas TF *tensors* represents the data fed into your model and the intermediate representations of that data as it passes through your model.

Initializing Variables in TensorFlow Graphs

Earlier in this chapter, you learned that variables are initialized in a TensorFlow session. Doing so will initialize variables with their computed values, such as the `lm` variable that is defined in terms of the variables `W` and `x`.

The TensorFlow method `tf.global_variables_initializer()` explicitly causes the initialization of all variables in your code, and it works in asynchronous mode.

TensorFlow Graph Execution

In all but trivial cases, a TF graph involves the following sequence of steps:

- Build (define) a TF graph
- Initialize a `tf.Session()`
- Feed data into the graph
- Execute the graph
- generate some output

In subsequent chapters, you will see TensorFlow code samples that illustrate each of the steps in the preceding list.

ARITHMETIC OPERATIONS IN TF GRAPHS

Listing 1.11 displays the contents of `tf-arithmetic.py` that illustrates how to perform arithmetic operations in a TF graph.

LISTING 1.11: tf-arithmetic.py

```
import tensorflow as tf

a = tf.add(4, 2)
b = tf.subtract(8, 6)
c = tf.multiply(a, 3)
```

```
d = tf.div(a, 6)

with tf.Session() as sess:
  print(sess.run(a)) # 6
  print(sess.run(b)) # 2
  print(sess.run(c)) # 18
  print(sess.run(d)) # 1
```

Listing 1.11 contains straightforward code for computing the sum, difference, product, and quotient via the `tf.add()`, `tf.subtract()`, `tf.multiply()`, and the `tf.div()` APIs, respectively. The `with` code block displays the result of invoking those APIs, and the output from Listing 1.11 is here:

```
6
2
18
1
```

TF GRAPHS AND BUILT-IN FUNCTIONS

Listing 1.12 displays the contents of `tf-math-ops.py` that illustrates how to perform additional arithmetic operations and how to use trigonometric functions in a TF graph.

LISTING 1.12: tf-math-ops.py

```
import tensorflow as tf

PI = 3.141592
sess = tf.Session()

print(sess.run(tf.div(12,8)))
print(sess.run(tf.floordiv(20.0,8.0)))
print(sess.run(tf.sin(PI)))
print(sess.run(tf.cos(PI)))
print(sess.run(tf.div(tf.sin(PI/4.), tf.cos(PI/4.))))
```

Listing 1.12 contains a hard-coded approximation for `PI`, an instance of `tf.Session()`, and four `print()` statements that display various arithmetic results. Note in particular the third output value is a very small number (the correct value is zero). The output from Listing 1.12 is here:

```
1
2.0
6.27833e-07
-1.0
1.0
```

Listing 1.13 displays the contents of `tf-math-ops-pi.py` that illustrates how to perform arithmetic operations and how to use `m.pi` for increased accuracy in a TF graph.

LISTING 1.13: tf-math-ops-pi.py

```
import tensorflow as tf
import math as m

PI = tf.constant(m.pi)

sess = tf.Session()
print(sess.run(tf.div(12,8)))
print(sess.run(tf.floordiv(20.0,8.0)))
print(sess.run(tf.sin(PI)))
print(sess.run(tf.cos(PI)))
print(sess.run(tf.div(tf.sin(PI/4.), tf.cos(PI/4.))))
```

Listing 1.13 contains the tf.floordiv(), tf.sin(), and tf.cos() APIs that calculate the floor of the quotient, the sine of a number (in radians), and the cosine of a number (in radians). This time the approximated value is one decimal place closer to the correct value of zero. The output from Listing 1.13 is here:

```
1
2.0
#OLD: 6.27833e-07
-8.742278e-08
-1.0
1.0
```

CALCULATING TRIGONOMETRIC VALUES IN TF

Listing 1.14 displays the contents of tf-trig-values.py that illustrates how to perform arithmetic operations in a TF graph.

LISTING 1.14: tf-trig-values.py

```
import tensorflow as tf

import numpy as np
import math as m
PI = tf.constant(m.pi)

a = tf.cos(PI/3.)
b = tf.sin(PI/3.)
c = 1.0/a # sec(60)
d = 1.0/tf.tan(PI/3.) # cot(60)

with tf.Session() as sess:
    print('a:', sess.run(a))
    print('b:', sess.run(b))
    print('c:', sess.run(c))
    print('d:', sess.run(d))
```

Listing 1.14 is straightforward: there are several of the same TF APIs that you saw in Listing 1.13. In addition, Listing 1.14 contains the `tf.tan()` API, which computes the tangent of a number (in radians). The output from Listing 1.14 is here:

```
a: 0.49999997
b: 0.86602545
c: 2.0000002
d: 0.57735026
```

CALCULATING EXPONENTIAL VALUES IN TF

Listing 1.15 displays the contents of `tf-exp-values.py` that illustrates how to perform exponential operations and how to use some TF built-in activation functions (discussed later) in a TF graph.

LISTING 1.15: tf-exp-values.py

```
import tensorflow as tf

a  = tf.exp(1.0)
b  = tf.exp(-2.0)
s1 = tf.sigmoid(2.0)
s2 = 1.0/(1.0 + b)
t2 = tf.tanh(2.0)

with tf.Session() as sess:
  print('a: ', sess.run(a))
  print('b: ', sess.run(b))
  print('s1:', sess.run(s1))
  print('s2:', sess.run(s2))
  print('t2:', sess.run(t2))
```

Listing 1.15 starts with the TF APIs `tf.exp()`, `tf.sigmoid()`, and `tf.tanh()` that compute the exponential value of a number, the sigmoid value of a number, and the hyperbolic tangent of a number, respectively. The second portion of Listing 1.15 is a with code block that displays the values of the initialized values. The output from Listing 1.15 is here:

```
a:  2.7182817
b:  0.13533528
s1: 0.880797
s2: 0.880797
t2: 0.9640276
```

The next section contains a simple example of using a for loop in TF, followed by another example that uses a for loop in TF.

USING FOR LOOPS IN TF

Listing 1.16 displays the contents of `tf-forloop1.py` that illustrates how to use a `for` loop in a TF graph.

LISTING 1.16: tf-forloop1.py

```
import tensorflow as tf

x = tf.Variable(0, name='x')

init = tf.global_variables_initializer()

with tf.Session() as session:
  session.run(init)

  for i in range(5):
    x = x + 1
    print(session.run(x))
```

Listing 1.16 initializes the TF variable x with the value 0 and then defines the `model` variable that is used for initializing the variables in this code sample (which is just the variable x). The `with` code block contains a loop that iterates through the values 1 through 5 inclusive. During each iteration of the loop, the variable x is incremented by 1 and its value is printed. The output from Listing 1.16 is here:

```
1
2
3
4
5
```

The next section contains an example of a `for` loop in TensorFlow using eager execution.

USING FOR LOOPS WITH EAGER EXECUTION IN TF

Listing 1.17 displays the contents of `tf-forloop2.py` that illustrates how to use a `for` loop in a TF graph.

LISTING 1.17: tf-forloop2.py

```
import tensorflow as tf

import tensorflow.contrib.eager as tfe
tfe.enable_eager_execution()

#x = tf.Variable(0, name='x')
x = tf.contrib.eager.Variable(0, name='x')

model = tf.global_variables_initializer()
```

```
for i in range(5):
    print(x)
    x = x + 1
```

Listing 1.17 contains code that is almost the same as Listing 1.16. The new code is shown in bold, starting with an `import` statement that enables us to enable eager execution (via the second statement in bold).

In addition, the definition of the TF variable is modified slightly to reference the `Variable` class in the `tf.contrib.eager` package instead of the `tf` package. The output from Listing 1.17 is here:

```
<tf.Variable 'x:0' shape=() dtype=int32, numpy=0>
tf.Tensor(1, shape=(), dtype=int32)
tf.Tensor(2, shape=(), dtype=int32)
tf.Tensor(3, shape=(), dtype=int32)
tf.Tensor(4, shape=(), dtype=int32)
```

USING WHILE LOOPS WITH EAGER EXECUTION IN TF

Listing 1.18 displays the contents of `tf-while-eager2.py` that illustrates how to use a `while` loop in a TF graph.

LISTING 1.18: tf-while-eager2.py

```
import tensorflow as tf
import tensorflow.contrib.eager as tfe

tfe.enable_eager_execution()

a = tf.constant(12)

while not tf.equal(a, 1):
    if tf.equal(a % 2, 0):
        a = a / 2
    else:
        a = 3 * a + 1
    print(a)
```

Listing 1.18 contains the required pair of statements to invoke eager execution, followed by the TF constant a of which the value is 12. The next portion of Listing 1.18 is a `while` loop that contains an `if/else` statement. If the value of a is even, then a is replaced by half its value. If a is odd, then its value is tripled and incremented by 1.

Did you notice that eager execution does not require a `with` code block in Listing 1.18? In addition, `sess.run()` statements are not required. Compared to the code samples that you have seen for deferred execution, eager execution involves a simpler syntax. The output from Listing 1.18 is here:

```
tf.Tensor(6.0, shape=(), dtype=float64)
tf.Tensor(3.0, shape=(), dtype=float64)
```

```
tf.Tensor(10.0,shape=(), dtype=float64)
tf.Tensor(5.0, shape=(), dtype=float64)
tf.Tensor(16.0,shape=(), dtype=float64)
tf.Tensor(8.0, shape=(), dtype=float64)
tf.Tensor(4.0, shape=(), dtype=float64)
tf.Tensor(2.0, shape=(), dtype=float64)
tf.Tensor(1.0, shape=(), dtype=float64)
```

As you can see, Listing 1.18 works correctly in eager mode because a is defined as a TF constant instead of a variable, so the issues in the previous code sample that contains a loop does not occur in this code sample.

THE TF.LESS() API IN A WHILE LOOP

Listing 1.19 displays the contents of tf-while-less.py that illustrates how to use a while loop in a TF graph.

LISTING 1.19: tf-while-less.py

```
import tensorflow as tf

x = tf.Variable(0., name='x')
threshold = tf.constant(5.)

model = tf.global_variables_initializer()

with tf.Session() as session:
  session.run(model)

  while session.run(tf.less(x, threshold)):
    x = x + 1
    x_value = session.run(x)
    print(x_value)
```

Listing 1.19 defines the TF variable x with the value 0, followed by the TF constant threshold, the value of which is 5. The variable model is defined as you have seen previously, followed by the with code block. Notice that the while loop inside the with code block has a more complex condition than the with code block that you saw previously in this chapter. The while loop increments x, assigns the result to x_value, and then prints the value of x_value as long as the value of x is less than the value of threshold. The output from Listing 1.19 is here:

```
1.0
2.0
3.0
4.0
5.0
```

As you can see, Listing 1.19 works correctly in eager mode.

THE TF ONE_HOT() API

Listing 1.20 displays the contents of `tf-onehot2.py` that illustrates how to use the TF `one_hot()` API with an array. Chapter 2 contains more information about "one hot" encoding, which you will encounter when you train neural networks.

LISTING 1.20: tf-onehot2.py

```
import tensorflow as tf

# Generate one-hot array using idx
idx = tf.get_variable("idx", initializer=tf.constant([2, 0,
                                                      -1, 0]))

target = tf.one_hot(idx, 3, 2, 0)

init = tf.global_variables_initializer()

with tf.Session() as sess:
  sess.run(init)
  print(sess.run(target))
```

Listing 1.20 starts by defining the variable `idx`, which is initialized as a TF constant that is a one-dimensional TF tensor with four integer values. Notice that the second and third elements in the TF tensor are equal, which means that their one-hot encoding will be the same. The next portion of Listing 1.20 defines `target`, which will contain the one-hot encoded values for `idx`. Next, the `with` code block initializes the TF variable `idx` and then prints the contents of `target`, which is a 4x3 tensor. The output from Listing 1.20 is here:

```
[[0 0 2]
 [2 0 0]
 [0 0 0]
 [2 0 0]]
```

ARRAYS IN TENSORFLOW (1)

Listing 1.21 displays the contents of `tf-elem1.py`, which illustrates how to define a TF array and access elements in that array.

LISTING 1.21: tf-elem1.py

```
import tensorflow as tf

sess = tf.Session()

arr1 = tf.constant([1,2])
print('arr1: ',sess.run(arr1))
print('[0]:  ',sess.run(arr1)[0])
print('[1]:  ',sess.run(arr1)[1])
```

Listing 1.21 contains the TF constant `arr1`, which is initialized with the value [1,2]. The three `print()` statements display the value of `arr1`, the value of the element, the index of which is 0, and the value of the element whose index is 1. The output from Listing 1.21 is here:

```
arr1:   [1 2]
[0]:    1
[1]:    2
```

ARRAYS IN TENSORFLOW (2)

Listing 1.22 displays the contents of `tf-elem2.py`, which illustrates how to define a TF array and access elements in that array.

LISTING 1.22: tf-elem2.py

```
import tensorflow as tf

sess = tf.Session()

arr2 = tf.constant([[1,2],[2,3]])
print('arr2:   ',sess.run(arr2))
print('[1]:    ',sess.run(arr2)[1])
print('[1,1]: ',sess.run(arr2)[1,1])
```

Listing 1.22 contains the TF constant `arr1` that is initialized with the value `[[1,2],[2,3]]`. The three `print()` statements display the value of `arr1`, the value of the element the index of which is 1, and the value of the element whose index is `[1,1]`. The output from Listing 1.22 is here:

```
arr2:   [[1 2]
 [2 3]]
[1]:    [2 3]
[1,1]:  3
```

ARRAYS IN TENSORFLOW (3)

Listing 1.23 displays the contents of `tf-elem3.py` that illustrates how to define a TF array and access elements in that array.

LISTING 1.23: tf-elem3.py

```
import tensorflow as tf

sess = tf.Session()

arr3 = tf.constant([[[1,2],[2,3]],[[3,4],[5,6]]])
print('arr3:   ',sess.run(arr3))
print('[1]:    ',sess.run(arr3)[1])
```

```
print('[1,1]:  ',sess.run(arr3)[1,1])
print('[1,1,0]:',sess.run(arr3)[1,1,0])
```

Listing 1.23 contains the TF constant `arr1` that is initialized with the value `[[[1,2],[2,3]],[[3,4],[5,6]]]`. The four `print()` statements display the value of `arr1`, the value of the element the index of which is 1, the value of the element whose index is `[1,1]`, and the value of the element whose index is `[1,1,0]`. The output from Listing 1.23 (adjusted slightly for display purposes) is here:

```
arr3:
 [[[1 2]
   [2 3]]

 [[3 4]
   [5 6]]]
[1]:
[[3 4]
 [5 6]]
[1,1]:   [5 6]
[1,1,0]: 5
```

MULTIPLYING TWO ARRAYS IN TF (CPU)

Listing 1.24 displays the contents of `tf-cpu.py` that illustrates how to define a TF array and access elements in that array.

LISTING 1.24: tf-cpu.py

```
import tensorflow as tf

with tf.Session() as sess:
    m1 = tf.constant([[3., 3.]])   # 1x2
    m2 = tf.constant([[2.],[2.]])  # 2x1
    p1 = tf.matmul(m1, m2)         # 1x1
    print('m1:',sess.run(m1))
    print('m2:',sess.run(m2))
    print('p1:',sess.run(p1))
```

Listing 1.24 contains two TF constants m1 and m2 that are initialized with the value `[[3., 3.]]` and `[[2.], [2.]]`, respectively. Due to the nested square brackets, m1 has shape 1x2, whereas m2 has shape 2x1. Hence, the product of m1 and m2 has shape `(1,1)`.

The three `print()` statements display the value of m1, m2, and p1. The output from Listing 1.24 is here:

```
m1: [[3. 3.]]
m2: [[2.]
 [2.]]
p1: [[12.]]
```

MULTIPLYING TWO ARRAYS IN TF (GPU)

Listing 1.25 displays the contents of tf-gpu.py, which illustrates how to define a TF array and access elements in that array.

LISTING 1.25: tf-gpu.py

```
import tensorflow as tf

with tf.Session() as sess:
  with tf.device("/gpu:1"):

  m1 = tf.constant([[3., 3.]])    # 1x2
  m2 = tf.constant([[2.],[2.]])   # 2x1
  p1 = tf.matmul(m1, m2)          # 1x1
  print('m1:',sess.run(m1))
  print('m2:',sess.run(m2))
  print('p1:',sess.run(p1))
```

Listing 1.25 contains the same code as Listing 1.24, with the exception of the second with code statement (shown in bold) that specifies the GPU where the TensorFlow code will be executed. Since the TensorFlow code is the same, the output is also the same. The only difference is that the code in Listing 1.25 is executed on a GPU instead of a CPU.

Note: Consider using Google Colaboratory that does not require specifying any GPUs in the code (a GPU is selected from a drop-down list), and also provides 12 free hours of GPU usage per day.

CONVERT PYTHON ARRAYS TO TF ARRAY

Listing 1.26 displays the contents of tf-convert-tensors.py that illustrates how to convert a Python array to a TF array.

LISTING 1.26: tf-convert-tensors.py

```
import tensorflow as tf
import numpy as np

x_data = np.array([[1.,2.],[3.,4.]])
x = tf.convert_to_tensor(x_data, dtype=tf.float32)

print ('x1:',x)
sess = tf.Session()
print('x2:',sess.run(x))
```

Listing 1.26 is straightforward, starting with an import statement for TensorFlow and one for NumPy. Next, the x_data variable is a NumPy array, and x is a TF tensor that results from converting x_data to a TF tensor. The output from Listing 1.26 is here:

```
x1: Tensor("Const:0", shape=(2, 2), dtype=float32)
x2: [[1. 2.]
 [3. 4.]]
```

WHAT IS TENSORBOARD?

TensorBoard is very powerful data and graph visualization tool that provides a great deal of useful information, as well as debugging support. Some of the previous code samples contain code snippets for saving TensorFlow graphs, and this section provides some additional information.

Listing 1.27 displays the contents of `tf-save-data.py`, which illustrates how to save a TF graph that can then be viewed in TensorBoard.

LISTING 1.27: tf-save-data.py

```
import tensorflow as tf

x = tf.constant(5,name="x")
y = tf.constant(8,name="y")
z = tf.Variable(2*x+3*y, name="z")
init = tf.global_variables_initializer()

with tf.Session() as session:
  writer = tf.summary.FileWriter("./tf_logs",session.graph)
  session.run(init)
  print('z = ',session.run(z)) # =>  z = 34
```

 # launch tensorboard with: tensorboard –logdir=./tf_logs

Listing 1.27 starts with code that you have seen in previous code samples. The second portion of Listing 1.27 contains a `with` code block that specifies the `tf_logs` subdirectory of the current directory as the location in which a TensorFlow graph will be saved. This subdirectory will be created if it does not already exist. Next, the TF variable `z` is initialized, and the result of that initialization is printed to standard output. In addition, the value of `z` is saved in an output file (located in the directory `tf_logs`) for the current TensorFlow graph. The output from Listing 1.27 is here:

34

Now open a command shell, navigate to the location of the TensorFlow code, and launch the following command to invoke TensorBoard (displayed in bold in Listing 1.27):

```
tensorboard -logdir=./tf_logs
```

Next, open a browser session and navigate to `localhost:6006`, and you will see the TensorFlow graph for the code in Listing 1.27.

Other tips and how-to information about TensorBoard are available here:
https://github.com/tensorflow/tensorboard/blob/master/README. md#my-tensorboard-isnt-showing-any-data-whats-wrong

GOOGLE COLABORATORY

GPU-based TensorFlow code is typically at least 15 times faster than CPU-based TensorFlow code. However, the cost of a good GPU can be a significant factor. Keep in mind that while NVIDIA provides GPUs, those consumer-based GPUs are not optimized for multi-GPU support (which *is* supported by TensorFlow).

Fortunately, Google Colaboratory is an affordable alternative that provides free GPU support, and also runs as a Jupyter notebook environment. In addition, Google Colaboratory executes your code in the cloud and involves zero configuration, and it is available here:

https://colab.research.google.com/notebooks/welcome.ipynb

This Jupyter notebook is suitable for training simple models and testing ideas quickly. Google Colaboratory makes it easy to upload local files, install software in Jupyter notebooks, and even connect Google Colaboratory to a Jupyter runtime on your local machine.

Some of the supported features of Colab include TensorFlow execution with GPUs, visualization using Matplotlib, and the ability to save a copy of your Google Colaboratory notebook to Github by using `File > Save a copy to GitHub`. Moreover, you can load any Jupyter notebook on GitHub by just adding the path to the URL `colab.research.google.com/github/` (see the website for details).

Google Colaboratory has support for other technologies, such as HTML and SVG, enabling you to render SVG-based graphics in notebooks that are in Google Colaboratory. One point to keep in mind: any software that you install in a Google Colaboratory notebook is only available on a per-session basis: if you log out and log in again, you need to perform the same installation steps that you performed during your earlier Google Colaboratory session.

As mentioned earlier, there is one other *very* nice feature of Google Colaboratory: you can execute code on a GPU for up to twelve hours per day for free. This free GPU support is extremely useful for people who do not have a suitable GPU on their local machine (which is probably the majority of users), and now they launch TensorFlow code to train neural networks in less than 20 or 30 minutes that would otherwise require multiple hours of CPU-based execution time.

Keep in mind the following details about Google Colaboratory. First, when you connect to a server in Google Colaboratory, you start what's known as a *session*. You can execute the code in a session with a CPU (the default), a GPU, or a TPU (which incurs a cost), and you can execute your code without any time limit for your session. However, if you select the GPU option for your session, *only the first 12 hours of GPU execution time are free during the session.* Any additional GPU time during that same session incurs a small charge (see the website for those details).

The other point to keep in mind is that any software that you install in a Jupyter notebook during a given session will *not* be saved when you exit that session. For example, the following code snippet installs `TFLearn` in a Jupyter notebook:

```
!pip install tflearn
```

When you exit the current session and at some point later you start a new session, you need to install `TFLearn` again, as well as any other software (such as Github repositories) that you also installed in any previous session.

OTHER CLOUD PLATFORMS

Google Cloud Platform (GCP) is a cloud-based service that enables you to train TensorFlow code in the cloud. GCP provides Deep Learning DL images (similar in concept to Amazon AMIs) that are available here:

https://cloud.google.com/deep-learning-vm/docs

The preceding link provides documentation, and also a link to DL images based on different technologies, including TensorFlow and PyTorch, with GPU and CPU versions of those images. Along with support for multiple versions of Python, you can work in a browser session or from the command line.

GCP SDK

Install GCloud SDK on a Mac-based laptop by downloading the software at this link:

https://cloud.google.com/sdk/docs/quickstart-macos

You will also receive USD 300 dollars worth of credit (over one year) if you have never used Google cloud.

SUMMARY

This chapter introduced you to TensorFlow, its architecture, and some of the tools that are part of the TensorFlow "family." Then you learned how to create a TensorFlow graph containing TensorFlow constants, placeholders, and variables. You also learned how to perform TensorFlow arithmetic operations, along with some built-in functions.

Next, you learned how to calculate trigonometric values, how to use for loops, while loops, and how to calculate exponential values. You also saw how to perform various operations on arrays, such as creating an identity matrix and a constant matrix.

You also learned how to save a TensorFlow graph and then view its contents in a browser using the built-in TensorBoard utility that is launched from the command line. Finally, you learned about Google Colaboratory, which is a Jupyter notebook-based environment that is accessible in a browser and executes your code in the cloud.

## CHAPTER 2

USEFUL TENSORFLOW APIS

This chapter focuses on useful TensorFlow APIs that you will encounter in many TensorFlow code samples. In fact, you will use these APIs in code samples that involve more complex TensorFlow code that are beyond the scope of this book.

An entire chapter devoted to APIs might seem like "dry" content, but the rationale is simple: you need *all* the APIs in this chapter if you continue learning about TensorFlow. Consequently, it's easier to find these TensorFlow APIs when they are located a single chapter. After you have read this chapter, you will be better prepared for the code samples that use these APIs in subsequent chapters.

Obviously, you are not "required" to read this chapter in its entirety, even if you are new to TensorFlow. However, it's worth skimming this chapter to make note of the TensorFlow APIs that are discussed. Doing so gives you a cursory overview of this chapter, and when you encounter code samples that use APIs in this chapter, you can read the appropriate section with details about those APIs.

The first part of this chapter briefly discusses some TensorFlow features, such as eager execution, which has a more Python-like syntax than "regular" TensorFlow syntax, and tensor operations (such as multiplying tensors). This section also shows you how to create for loops and while loops in TensorFlow. Recall from Chapter 1 that TensorFlow 2 uses eager execution as the default execution whereas Tensorflow 1.x uses deferred execution. As this book goes to print, TensorFlow 2 has not been released, which means that the code samples in this book are for TensorFlow 1.x.

The second part of this chapter contains a collection of TensorFlow code samples that show you how to use various APIs that are commonly used in machine learning. Specifically, you will see how to use the `tf.random_normal()` API for generating random numbers (which is useful for initializing the weights of edges in neural networks); the `tf.argmax()` API for finding the index of each row (or column) that contains the maximum value in each row (or column), which is used for calculating the accuracy of the training process involving various algorithms; and also the `tf.range()` API, which is similar to the `np.linspace()` API that you saw in Chapter 1.

The third portion of this chapter discusses another set of TensorFlow APIs, including `reduce_mean()` and `equal()`, both of which are involved in calculating the accuracy of a the training of a neural network (in conjunction with `tf.argmax()`). You will also learn about the `truncated_normal()` API, which is a variant of the `tf.random_normal()` API, and the `one_hot()` API for encoding data in a particular fashion (i.e., the digit 1 in one position and zero in all other positions of a vector). One of the most frequently used APIs is `reshape()`, which you will see in any TensorFlow code that involves training a Convolutional Neural Network (CNN). After you have completed this section of the chapter, navigate to the following URL that contains a massive collection of TF APIs:

https://www.tensorflow.org/api_docs/python/tf

The fourth portion of this chapter offers more details about TensorBoard (which was briefly introduced in Chapter 1), and shows you how to use TensorBoard APIs in TensorFlow code in order to "augment" a TensorFlow graph with supplemental information that can be rendered in TensorBoard in a Web browser.

TF TENSOR OPERATIONS

TF supports many arithmetic operations on TensorFlow tensors, such as:

Adding pairs of tensors
Multiplying tensors
Dividing tensors
Subtracting tensors

The preceding operations are performed on an element-by-element basis of two tensors. For example, adding two 2x2 tensors involves four additions, whereas adding two 4x4 tensors involves sixteen additions.

The TF `tf.argmax()` API enables you to find the maximum value of each row (or each column) of a two-dimensional TF tensor. This API is used in CNNs as part of the calculation of the number of images (in the case of MNIST) that are correctly identified during the training phase. The TF `tf.argmin()` API is similar to the `tf.argmax()` API, except that minimum values are found instead of maximum values.

TensorFlow provides statistical methods, such as `tf.reduce_mean()` and `tf.random_normal()`, for calculating the mean of a set of numbers and randomly selecting numbers from a normal distribution. The `tf.truncated_normal()` API is similar to the `tf.random_normal()` API, with the added constraint that the selected numbers must be in a specified range (which is specified by you).

Other useful TensorFlow APIs are shown in the following list:

eval()
rank()
reduce_sum()
reshape()
range()
equal()
one_hot()
name_scope()
summary.scalar()

Later in this chapter, you will see code samples that illustrate how to use most of the TF APIs in the preceding list.

TENSORFLOW AND BUFFERS

Listing 2.1 displays the contents of `tf-basic-buffer.py`, which illustrates how to save the contents of a TF graph to a TF buffer.

LISTING 2.1: tf-basic-buffer.py

```
import tensorflow as tf

g = tf.Graph()

with g.as_default():
  sess = tf.Session()
  a = tf.placeholder("float", name="a")
  b = tf.placeholder("float", name="b")
  c = tf.multiply(a,b, name="c")

  feed_dict = {a:2, b:3}
  print(sess.run(c, feed_dict))
  print (g.as_graph_def())
```

Listing 2.1 contains the standard `import` statement and then initializes the variable g with the current (default) TensorFlow graph. Next, the `with` code block defines the TF placeholders a and b, followed by the TF variable c that is the product of a and b. Next, `feed_dict` is initialized with the values 2 and 3 for a and b, respectively, thereby enabling the next `print()` statement to print the value of c. A portion of the output from launching the code in Listing 2.1 is here:

```
6.0
node {
  name: "a"
```

```
op: "Placeholder"
attr {
  key: "dtype"
  value {
    type: DT_FLOAT
  }
}
attr {
  key: "shape"
  value {
    shape {
      unknown_rank: true
    }
  }
}
}
// node "b" omitted for brevity
node {
name: "c"
op: "Mul"
input: "a"
input: "b"
attr {
  key: "T"
  value {
    type: DT_FLOAT
  }
}
}
```

Later in this chapter you will learn how to use TensorBoard in order to display the graph that corresponds to TensorFlow code (such as Listing 2.1).

TF OPERATIONS WITH RANDOM NUMBERS

TF provides APIs for generating random numbers, such as the TF `tf.random_normal()` API that generates random numbers from a normal distribution. Listing 2.2 displays the contents of `tf-normal-dist.py` that illustrates how to use the `tf.random_normal()` method in a TF graph.

LISTING 2.2: tf-normal-dist.py

```python
import tensorflow as tf

# normal distribution:
w = tf.Variable(tf.random_normal([784, 10], stddev=0.01))

# mean of an array:
b = tf.Variable([10,20,30,40,50,60],name='t')

with tf.Session() as sess:
  sess.run(tf.global_variables_initializer())
  print("w: ",sess.run(w))
  print("b: ",sess.run(tf.reduce_mean(b)))
```

Listing 2.2 defines the TF variables w (initialized with random values) and b (initialized with hard-coded values). The TF w variable has dimensions 784x10, whose rows correspond to a "flattened" 28x28 image, and the columns correspond to the 10 digits 0 through 9. You will frequently see TF tensors with shape 784x10 when you work with images in CNNs in TensorFlow (and Keras as well).

```
w:  [[-1.1800987e-03 -5.4473975e-03  9.7551057e-03 ...
                                              -8.1719309e-03
    1.7652005e-02 -5.8396067e-04]
 [ 2.4753861e-04 -2.3292618e-02  9.7491527e-03 ...
                                               5.6430311e-03
    6.7094434e-03 -3.2657082e-04]
 [-7.6869861e-03  1.1552800e-02  1.0761258e-02 ...
                                              -3.4506479e-03
    5.9720408e-03  3.5639866e-03]
 ...
 [ 1.6964199e-02 -1.0729434e-02  2.0073706e-02 ...
                                              -1.3808235e-02
   -1.0414679e-03  1.2187610e-02]
 [-2.9807910e-03  1.3769536e-03 -8.6986246e-03 ...
                                               7.6331957e-03
    7.1591261e-04 -2.0289846e-02]
 [-5.4767268e-04 -1.2888687e-02  5.1014713e-04 ...
                                              -1.1962975e-02
   -7.6868264e-03  9.6106824e-06]]
b:   35
```

Listing 2.3 displays the contents of tf-random-normal.py that illustrates how to use the tf.random_normal() method in a TF graph.

LISTING 2.3: *tf-random-normal.py*

```
import tensorflow as tf

# initialize a 6x3 array of random numbers:
values = {'weights':tf.Variable(tf.random_normal([6,3]))}

with tf.Session() as sess:
  sess.run(tf.global_variables_initializer())
  print(sess.run(values['weights']))
```

Listing 2.3 initializes the values variable with the element weights, which is initialized as a TF variable that comprises a 6x3 tensor containing randomly selected values from a normal distribution. The output from launching the code in Listing 2.3 is here:

```
[[ 0.83530873  1.00292623 -0.11506019]
 [-0.82566637  0.45110381 -1.09818149]
 [-0.26840609 -0.4434818   0.48984894]
 [ 0.48618484 -1.55438101  1.38592315]
 [ 0.23166555 -0.37035158  0.10339094]
 [-0.58803022 -0.17280568 -0.96742302] ]
```

TF ARRAYS AND MAXIMUM VALUES

The TF `tf.argmax()` API determines the index values containing maximum values on a row-wise basis or on a column-wise basis for a TF tensor. Just to be sure you understand the previous statement: the TF `tf.argmax()` API determines the *index* values that contain maximum values and *not* the actual maximum values in those index positions. As a trivial example, the array [10,20,30] contains a minimum value of 10 in index position 0 and a maximum value of 30 in index position 2. Consequently, TF `tf.argmax()` API returns the value 2, whereas the TF `tf.argmin()` API returns the value 0.

Listing 2.4 displays the contents of `tf-row-max.py` that illustrates how to find the maximum value on a row-wise basis in a TF array.

LISTING 2.4: tf-row-max.py

```
import tensorflow as tf

# initialize an array of arrays:
a = [[1,2,3], [30,20,10], [40,60,50]]
b = tf.Variable(a, name='b')

with tf.Session() as sess:
  sess.run(tf.global_variables_initializer())
  print("b: ",sess.run(tf.argmax(b,1)))
```

Listing 2.4 defines the Python variable a as a 3x3 array of integers. Next, the variable b is initialized as a TF variable that is based on the contents of the Python variable a. The `with` statement simply displays the value of b, which is shown here:

```
[ 1.3  1.   4.   23.5]
1.3
4.0
```

Notice that `tf.argmax()` in Listing 2.4 specifies the value 1 (shown in bold): this indicates that you want the indexes containing *row-wise* maximum values. On the other hand, specify the value 0 if you want the indexes containing *column-wise* maximum values.

TF RANGE() API

Listing 2.5 displays the contents of `tf-range1.py` that illustrates how to use the TF `range()` API, which generates a range of numbers between an initial value and a final value, where consecutive values differ by the same constant.

LISTING 2.5: tf-range1.py

```
import tensorflow as tf

a1 = tf.range(3, 18, 3)
a2 = tf.range(0, 8, 2)
a3 = tf.range(-6, 6, 3)
a4 = tf.range(-10, 10, 4)

with tf.Session() as sess:
  print('a1:',sess.run(a1))
  print('a2:',sess.run(a2))
  print('a3:',sess.run(a3))
  print('a4:',sess.run(a4))
```

Listing 2.5 defines the TF variable a1 that is the set of numbers between 3 (inclusive) and 18 (exclusive), where each number is 3 larger than its predecessor. Similarly, the variables a2, a3, and a4 are defined, with ranges that have a different start value, and value, and increment value. The output from launching the code in tf-range1.py is here:

```
('a1:', array([3,  6,  9,  12,  15], dtype=int32))
('a2:', array([0,  2,  4,  6], dtype=int32))
('a3:', array([-6,  -3,   0,   3], dtype=int32))
('a4:', array([-10,  -6,  -2,  2,  6], dtype=int32))
```

OPERATIONS WITH NODES (1)

Listing 2.6 displays the contents of tf-addnodes1.py, which illustrates how to add two nodes (which happen to be TF placeholders) in a TF graph.

LISTING 2.6: tf-addnodes1.py

```
import tensorflow as tf

a1 = tf.placeholder(tf.float32)
a2 = tf.placeholder(tf.float32)
a3 = a1 + a2

sess = tf.Session()
print(sess.run(a3, {a1:7, a2:13}))
print(sess.run(a3, {a1:[7,10], a2:[13,20]}))
```

Listing 2.6 defines the TF placeholders a1, a2, and a3, where a3 is defined as the sum of a1 and a2. Next, the TF variable sess is instantiated as an instance of the tf.Session() class, followed by two print() statements. Notice that both print() statements display the value of the variable a3 by specifying different values for a1 and a2. The output from launching the code in Listing 2.6 is here:

```
20.0
[20.  30.]
```

OPERATIONS WITH NODES (2)

Listing 2.7 displays the contents of tf-addnodes2.py, which illustrates another way to add two nodes in a TF graph.

LISTING 2.7: tf-addnodes2.py

```
import tensorflow as tf

a1 = tf.placeholder(tf.float32)
a2 = tf.placeholder(tf.float32)
a3 = a1 + a2
a4 = a3*6
a5 = a4/2

sess = tf.Session()
print(sess.run(a4, {a1:7, a2:13}))
print(sess.run(a5, {a1:[7,10], a2:[13,20]}))
```

Listing 2.7 performs numeric calculations by defining the TF placeholders a1, a2, a3, a4, and a5, where a3 is the sum of a1 and a2, a4 is the product of a3 and the number 6, and a5 is the value of a4 divided by 2. Next, the TF variable sess is instantiated as an instance of the tf.Session() class, followed by two print() statements. The first print() statement displays the value of the variable a4 by specifying scalar values for a1 and a2, whereas the second print() statement displays the value of the variable a5 by specifying scalar value for a1 and a one-dimensional array for the value of a2. The output from launching the code in Listing 2.7 is here:

```
120.0
[60.  90.]
```

OPERATIONS WITH NODES (3)

Listing 2.8 displays the contents of tf-addnodes3.py that illustrates a third way to add two nodes in a TF graph.

LISTING 2.8: tf-addnodes3.py

```
import tensorflow as tf

a1 = tf.constant(5,tf.float32,name='a1')
```

```
a2 = tf.constant(9,tf.float32,name='a2')
a3 = a1 + a2
a4 = a3*6
a5 = a4/2

with tf.Session() as sess:
  print('a4:',sess.run(a4))
  print('a5:',sess.run(a5))

  b1 = tf.add_n([a4, a5], name="b1")
  print('b1:',sess.run(b1))

  b2 = tf.multiply(a4, a5, name="b2")
  print('b2:',sess.run(b2))

  b3 = tf.multiply(tf.pow(a4,2), tf.pow(a5,2), name="b3")
  print('b3:',sess.run(b3))
```

Listing 2.8 contains code that is similar to the code samples in Listing 2.7 and Listing 2.6. In this example, the variables b1, b2, and b3 are initialized inside the with code block. Specifically, b1 is sum of a4 and a5, b2 is the product of a4 and a5, and b3 is the product of a4 squared and a5 squared. The output from launching the code in Listing 2.8 is here:

```
a4: 84.0
a5: 42.0
b1: 126.0
b2: 3528.0
b3: 12446784.0
```

THE TF.EVAL() API

Listing 2.9 displays the contents of tf-eval.py that illustrates how to invoke the TF eval() API that provides another mechanism for evaluating TF tensors.

LISTING 2.9: tf-eval.py

```
import tensorflow as tf

a = tf.constant([8], tf.int32, name="a")
x = tf.placeholder(tf.int32, name="x")

y = a * x
with tf.Session() as sess:
  print('y:',y.eval(feed_dict={x:[3]}))
```

Listing 2.9 initializes the TF constant a and declares the TF placeholder x. Next, the variable y is defined as the product of a and x, followed by a with code block that uses the TF eval() method to calculate the value of y. Notice that `feed_dict` is required in order to specify a value for the TF placeholder x. The output from launching the code in Listing 2.9 is here:

```
y: [24]
```

THE TF.SIZE(), TF.SHAPE(), AND TF.RANK() APIS

These three TensorFlow APIs are somewhat related to each other, so they are included in the same section for easy reference.

The `tf.size()` API returns the number of elements in a TF tensor. Here is a simple example:

```
t = tf.constant([[[1,1,1],[2,2,2]],[[3,3,3],[4,4,4]]])
tf.size(t)   # 12
```

In essence, ignore the square brackets and count the number of elements in order to determine the answer.

The `tf.shape()` API returns the shape of a TF tensor, which is the number of elements in each "dimension." Here is a simple example:

```
t = tf.constant([[[1,1,1],[2,2,2]],[[3,3,3],[4,4,4]]])
tf.shape(t)   # [2, 2, 3]
```

The `tf.rank()` API returns the number of indices required uniquely to select each element of the tensor. The rank is also known as the "order," "degree" or "ndims." Here is a simple example:

```
# the shape of tensor 't' is [2, 2, 3]
t = tf.constant([[[1,1,1],[2,2,2]],[[3,3,3],[4,4,4]]])
# the rank of t is 3
```

Note that the rank of a tensor does not equal the rank of a matrix.

You might initially think that the rank in the preceding example is 4 instead of 3 because it's easy to overlook the number of nested square brackets. An easier way to display the preceding TF tensor is shown here:

```
[
  [
    [1,1,1],
    [2,2,2]
  ],
  [
    [3,3,3],
    [4,4,4]
  ]
]
```

As you can see in the preceding layout, the rank of the tensor is 3 because you need to "traverse" 3 levels in order uniquely to identify each element of the tensor.

Listing 2.10 displays the contents of `tf-rank.py`, which illustrates how to invoke the TF `rank()` API.

LISTING 2.10: tf-rank.py

```
import tensorflow as tf

b1 = tf.constant(7)
b2 = tf.constant([3,7])
b3 = tf.constant([[3,7],[11,13]])

sess = tf.Session()
print(sess.run(tf.rank(b1)))
print(sess.run(tf.rank(b2)))
print(sess.run(tf.rank(b3)))
```

Listing 2.10 is a simple example that defines the TF constants b1, b2, and b3 that have dimensions 0, 1, and 2, respectively. The three `print()` statements display the rank of these TF constants, which equals their dimensionality. The output from launching the code in Listing 2.10 is here:

```
0
1
2
```

If you know the shape of a TF tensor, you can determine its rank and its size. In the examples above, the shape is [2,2,3] and therefore the rank is 3 (which equals the number of elements in [2,2,3]) and the size is 12 (2x2x3 = 12).

THE TF.REDUCE_PROD() AND TF.REDUCE_SUM() API

Listing 2.11 displays the contents of `tf-reduce-prod.py`, which illustrates how to invoke the TF `reduce_prod()` and `reduce_sum()` APIs for multiplying and adding, respectively, the numeric elements in a TF tensor.

LISTING 2.11: tf-reduce-prod.py

```
import tensorflow as tf

x = tf.constant([100,200,300], name="x")
y = tf.constant([1,2,3], name="y")

sum_x  = tf.reduce_sum(x, name="sum_x")
prod_y = tf.reduce_prod(y, name="prod_y")
div_xy = tf.div(sum_x, prod_y, name="div_xy")

sess = tf.Session()
print(sess.run(sum_x))
print(sess.run(prod_y))
print(sess.run(div_xy))
sess.close()
```

Listing 2.11 defines the TF constants x and y, followed by three variables of which the values are based on three TF APIs. Specifically, sum_x equals the value of invoking the tf.reduce_sum() API with the TF constant x, which equals the sum of the numeric elements of x. Next, prod_y equals the value of invoking the tf.reduce_prod() API with the TF constant y, which equals the product of the numeric elements of y. Finally, div_xy equals the ratio of sum_x and prod_y. The output from launching the code in Listing 2.11 is here:

```
600
6
100
```

THE TF.REDUCE_MEAN() API

Listing 2.12 displays the contents of tf-reduce-mean.py that illustrates how to invoke the TF reduce_mean() APIs.

LISTING 2.12: tf-reduce-mean.py

```
import tensorflow as tf

x = tf.constant([100,200,300], name='x')
y = tf.constant([1,2,3], name='y')

sum_x  = tf.reduce_sum(x, name="sum_x")
prod_y = tf.reduce_prod(y, name="prod_y")
mean   = tf.reduce_mean([sum_x,prod_y], name="mean")

sess = tf.Session()
print(sess.run(mean))
sess.close()
```

Listing 2.12 defines the TF constants x and y, followed by three variables, the values of which are based on three TF APIs. Specifically, sum_x equals the value of invoking the tf.reduce_sum() API with the TF constant x, which equals the sum of the numeric elements of x. Next, prod_y equals the value of invoking the tf.reduce_prod() API with the TF constant y, which equals the product of the numeric elements of y. Finally, mean equals the sum of sum_x and prod_y. The output from launching the code in Listing 2.12 is here:

```
303
```

THE TF.REDUCE_MEAN() API

Listing 2.13 displays the contents of tf-mean1.py, which illustrates how to invoke the TF tf.reduce_mean() APIs.

LISTING 2.13: tf-reduce-mean2.py

```
import tensorflow as tf

x1 = tf.constant([[1,1,1],[2,2,2],[3,3,3],[4,4,4]],
                                        dtype=tf.float32)


y1 = tf.reduce_mean(x1, [0])
y2 = tf.reduce_mean(x1, [0,1])

sess = tf.Session()
print("y1:",sess.run(y1))
print("y2:",sess.run(y2))
sess.close()
```

Listing 2.13 defines the TF constant x1, which is a 4x3 tensor. The next portion of Listing 2.13 defines y1 and y2 via the TF API tf.reduce_sum(). The contents of y1 is the column-wise averages, and the contents of y2 is the row-wise averages. The output from Listing 2.13 is here:

```
('y1:', array([2.5, 2.5, 2.5], dtype=float32))
('y2:', 2.5)
```

THE TF RANDOM_NORMAL() API (2)

The TensorFlow tf.random_normal() API returns a set of values from a normal distribution with mean equal to 0 and standard deviation equal to 1 as the default values.

Listing 2.14 displays the contents of tf-rand-normal2.py that illustrates how to use the random_normal() API with a NumPy array.

LISTING 2.14: tf-rand-normal2.py

```
import tensorflow as tf

init = tf.global_variables_initializer()

with tf.Session() as session:
  session.run(init)

  for i in range(5):
    x_train = tf.random_normal((1,), mean=5, stddev=2.0)
    y_train = x_train * 2 + 3
    print("x_train:",x_train.eval())

  for i in range(5):
    x_train = tf.random_normal((2,), mean=5, stddev=2.0)
    y_train = x_train * 2 + 4
    print("x_train:",x_train.eval())

  for i in range(5):
    x_train = tf.random_normal((3,), mean=5, stddev=2.0)
```

```
y_train = x_train * 2 + 6
print("x_train:",x_train.eval())
```

Listing 2.14 contains three `for` loops, all of which initialize the variable `x_train` by invoking the `tf.random_normal()` API with the mean equal to 5 and the `stddev` equal to 2.0. Moreover, all three loops define the `y_train` variable as a linear combination of the `x_train` values. The first parameter of the `tf.random_normal()` API specifies the shape of the set of random numbers. This parameter is set to `(1,)`, `(2,)`, and `(3,)` in the three `for` loops, which means that there will be one, two, and three columns of output, respectively.

The output from Listing 2.14 displays the values of the `x_train` variable in each of the three `for` loops, as shown here:

```
x_train: [5.2466145]
x_train: [3.7596474]
x_train: [5.380934]
x_train: [4.9017477]
x_train: [4.3671246]
x_train: [3.786654  4.2114644]
x_train: [3.651453  0.66866636]
x_train: [7.1243105 4.9335704]
x_train: [8.221105  3.190276]
x_train: [1.2022083 7.333288 ]
x_train: [5.0063868 8.611422   5.285639 ]
x_train: [8.9014015 8.133568   5.4956264]
x_train: [3.7392507 6.370685   4.959084 ]
x_train: [ 2.477441 -0.8832741 6.595917 ]
x_train: [5.823818  5.154063   7.543862]
```

THE TF.TRUNCATED_NORMAL() API

The `tf.truncated_normal()` API produces a set of random values from a *truncated* normal distribution, which differs from the `tf.random_normal()` API in terms of the interval from which random values are selected. First, visualize a regular normal distribution the mean of which is close to 0. Second, mentally "chop off" values that are more than 2 standard deviations from the mean, which results in a "truncated" normal distribution.

This truncated interval is the interval from which random numbers are selected. Specifically, randomly generate a number and include that number in the "result" set if it's *inside* the "truncated" normal distribution. However, if the randomly chosen value lies *outside* the truncated normal distribution, regenerate the value (and do so repeatedly if it's necessary) until it's *inside* the truncated normal distribution, after which the number is included in the "result" set.

The `tf.truncated_normal()` API helps to prevent (or at least reduce) saturation that can occur with the sigmoid function: neurons stop "learning" if saturation occurs.

THE TF.RESHAPE() API

Listing 2.15 displays the contents of `tf-reshape.py`, which illustrates how to invoke the TF `reshape()` APIs in order to create TF tensors with different shapes.

LISTING 2.15: tf-reshape.py

```
import tensorflow as tf

x = tf.constant([[2,5,3,-5],[0,3,-2,5],[4,3,5,3]])

sess = tf.Session()
print(sess.run(tf.shape(x)))
print('1:',sess.run(tf.reshape(x, [6,2])))

with sess.as_default():
  print('2:',tf.reshape(x, [3,4]).eval())
```

Listing 2.15 defines the TF constant x as a TF tensor with shape (3,4) that consists of 12 integers (some are positive and some are negative). We can reshape the variable x as long as the new row size and column size have a product value of 12.

Hence, the allowable pairs of values for rows and columns are: 1 and 12, 2 and 6, or 3 and 4 (which are the dimensions of the TF variable x). The output from launching the code in Listing 2.20 is here:

```
('1:', array([[ 2,   5],
              [ 3,  -5],
              [ 0,   3],
              [-2,   5],
              [ 4,   3],
              [ 5,   3]], dtype=int32))

('2:', array([
       [ 2,   5,   3,  -5],
       [ 0,   3,  -2,   5],
       [ 4,   3,   5,   3]], dtype=int32))
```

THE TF.EQUAL() API (1)

Listing 2.16 displays the contents of `tf-equal1.py`, which illustrates how to invoke the TF `equal()` API as well as the TF `not_equal()` API to determine whether or not two TF tensors are equal.

LISTING 2.16: tf-equal1.py

```
import tensorflow as tf

x1  = tf.constant([0.9, 2.5, 2.3, -4.5])
x2  = tf.constant([1.0, 2.0, 2.0, -4.0])
eq  = tf.equal(x1,x2)
neq = tf.not_equal(x1,x2)

with tf.Session() as sess:
  print('x1: ',sess.run(x1))
  print('x2: ',sess.run(x2))
  print('eq: ',sess.run(eq))
  print('neq:',sess.run(neq))
```

Listing 2.16 defines the TF constants x1 and x2 as one-dimensional constants. Next, the variable eq is defined by performing an element-by-element comparison of x1 and x2, and the result of the comparison is a one-dimensional array of Boolean values. The output from launching the code in Listing 2.16 is here:

```
x1:  [ 0.9  2.5  2.3 -4.5]
x2:  [ 1.   2.   2.  -4.]
eq:  [False False False False]
neq: [ True  True  True  True]
```

THE TF.EQUAL() API (2)

Listing 2.17 displays the contents of tf-equal2.py, which also illustrates how to invoke the TF equal() API.

LISTING 2.17: tf-equal2.py

```
import tensorflow as tf

import numpy as np

x1 = tf.constant([0.9, 2.5, 2.3, -4.5])
x2 = tf.constant([1.0, 2.0, 2.0, -4.0])
x3 = tf.Variable(x1)

with tf.Session() as sess:
  sess.run(tf.global_variables_initializer())
  print('x1:',sess.run(x1))
  print('x2:',sess.run(x2))
  print('r3:',sess.run(tf.round(x3)))
  print('eq:',sess.run(tf.equal(x1,x3)))
```

Listing 2.17 defines the TF constants x1, x2, and x3, which contain an assortment of positive and negative decimal values. The with code block displays the values of x1 and x2, along with r3 (which rounds the values in x3). The fourth print() statement generates an array of Boolean values that are based on an element-by-element comparison of the elements of x1 and x3. The output from launching the code in Listing 2.17 is here:

```
x1: [ 0.9   2.5   2.3  -4.5]
x2: [ 1.    2.    2.   -4.]
r3: [ 1.    2.    2.   -4.]
eq: [ True  True  True  True]
```

THE TF.ARGMAX() API (1)

Listing 2.18 displays the contents of tf-argmax.py, which illustrates how to invoke the TF argmax() API.

LISTING 2.18: tf-argmax1.py

```
import tensorflow as tf

import numpy as np

x1 = tf.constant([3.9, 2.1, 2.3, -4.0])
x2 = tf.constant([1.0, 2.0, 5.0, -4.2])

with tf.Session() as sess:
  sess.run(tf.global_variables_initializer())
  print('x1:',sess.run(x1))
  print('x2:',sess.run(x2))

  print('a1:',sess.run(tf.argmax(x1, 0)))
  print('a2:',sess.run(tf.argmax(x2, 0)))
```

Listing 2.18 defines the TF constants x1 and x2, which are one-dimensional tensors that contain positive and negative decimal values. The with code block displays the values of x1 and x2, followed by the indexes where the maximum values appear in x1 and in x2. The output from launching the code in Listing 2.18 is here:

```
x1: [ 3.9   2.1   2.3  -4.  ]
x2: [ 1.    2.    5.   -4.2]
a1: 0
a2: 2
```

THE TF.ARGMAX() API (2)

Listing 2.19 displays the contents of tf-argmax2.py, which illustrates how to invoke the TF argmax() API.

LISTING 2.19: tf-argmax2.py

```
import tensorflow as tf
import numpy as np
```

```
# initialize array of arrays:
a = [[1,2,3], [30,20,10], [40,60,50]]
b = tf.Variable(a, name='b')

with tf.Session() as sess:
  sess.run(tf.global_variables_initializer())
  print("index of max values in b: ",
        sess.run(tf.argmax(b,1)))
```

Listing 2.19 defines the 3x3 array a that contains integer values, followed by the definition of the TF variable b. The `with` code block displays the index of each row of a that contains the maximum value for that row. The output from launching the code in Listing 2.19 is here:

```
index of max values in b:   [2 0 1]
```

THE TF.ARGMAX() API (3)

Listing 2.20 displays the contents of `tf-argmax3.py` with another example of invoking the TF `argmax()` API.

LISTING 2.20: tf-argmax3.py

```
import tensorflow as tf
import numpy as np

x = np.array([[31, 23,  4, 54],
              [18,  3, 25,  0],
              [28, 14, 33, 22],
              [17, 12,  5, 81]])

y = np.array([[31, 23,  4, 24],
              [18,  3, 25,  0],
              [28, 14, 33, 22],
              [17, 12,  5, 11]])

with tf.Session() as sess:
  sess.run(tf.global_variables_initializer())
  print('xmax:', sess.run(tf.argmax(x,1)))
  print('ymax:', sess.run(tf.argmax(y,1)))
  print('equal:',sess.run(tf.equal(x,y)))
```

Listing 2.20 defines the 3x3 NumPy arrays x and y, which contain integer values. The `with` code block contains a `print()` statement that displays the index of the maximum value for each *row* of x, followed by another `print()` statement that displays the index of the maximum value for each row of y. The third `print()` statement displays an array of Boolean values that are the result of performing an element-by-element comparison of the elements of x and y to determine which pairs contain are equal value. The output from launching the code in Listing 2.20 is here:

```
xmax: [3 2 2 3]
ymax: [0 2 2 0]
equal: [[ True   True   True False]
 [ True   True   True   True]
 [ True   True   True   True]
 [ True   True   True False]]
```

COMBINING TF.ARGMAX() AND TF.EQUAL() APIS

Listing 2.21 displays the contents of tf-argmax-equal1.py, which illustrates how to invoke the TF equal() API with the TF tf.argmax() API.

LISTING 2.21: tf-argmax-equal1.py

```
import tensorflow as tf
import numpy as np

pred = np.array([[31,   23,   4, 24, 27, 34],
                 [18,    3,  25,  0,  6, 35],
                 [28,   14,  33, 22, 20,  8],
                 [13,   30,  21, 19,  7,  9],
                 [16,    1,  26, 32,  2, 29],
                 [17,   12,   5, 11, 10, 15]])

y =      np.array([[31,   23,   4, 24, 27, 14],
                   [18,    3,  25,  0,  6, 35],
                   [28,   14,  33, 22, 20,  8],
                   [13,   30,  21, 19,  7,  9],
                   [16,    1,  26, 32,  2, 29],
                   [17,   12,   5, 11, 10, 15]])

with tf.Session() as sess:
  correct_pred = tf.equal(tf.argmax(pred, 1),tf.argmax(y,
                                                        1))
  accuracy = tf.reduce_mean(tf.cast(correct_pred,
                                                tf.float32))

  print("correct_pred:",sess.run(correct_pred))
  print("accuracy:",sess.run(accuracy))
```

Listing 2.21 defines two NumPy two-dimensional arrays of integers. The purpose of the with code block is to determine the indexes of the maximum row values of x and y, and then compare those index values to determine how often they are equal. The result (after multiplying by 100) gives us the percentage of occurrences of equal index positions.

In Listing 2.21, the maximum value in each of the rows 2 through 6 of x are in the same position as the maximum value for rows 2 through 6 of y. The index of the maximum value in row 1 of x is 5, whereas the index of the maximum value in row 1 of y is 0 (so the index values do not match). Hence, the index values match in 5 of the 6 rows, which is 0.8333333 (rounded to six decimal places), and that is the decimal value for 83.33333%.

This code sample is very helpful for understanding the logic (which is the same as this code sample) to calculate the accuracy of the training and testing portion of CNNs that are trained for the purpose of correctly identifying images. The output from launching the code in Listing 2.27 is here:

```
correct_pred: [False  True  True  True  True  True]
accuracy: 0.8333333
```

COMBINING TF.ARGMAX() AND TF.EQUAL() APIS

Listing 2.22 displays the contents of tf-argmax-equal2.py that illustrates how to invoke the TF equal() API with the TF argmax() API.

LISTING 2.22: tf-argmax-equal2.py

```
import tensorflow as tf
import numpy as np

# cat, dog, bird, fish, carrot, apple
# predictions from our model:
pred = np.array([[0.1, 0.03, 0.2, 0.05, 0.02, 0.6],
                 [0.5, 0.04, 0.2, 0.06, 0.10, 0.1],
                 [0.2, 0.04, 0.5, 0.06, 0.10, 0.1]])

# true values from our labeled data:
y_vals = np.array([[0,  0,  0,  0,  0,  1],
                   [1,  0,  0,  0,  0,  0],
                   [0,  0,  1,  0,  0,  0]])

with tf.Session() as sess:
  print("argmax(pred,1):   ", sess.run(tf.argmax(pred,1)))
  print("argmax(y_vals,1):", sess.run(tf.argmax(y_vals,1)))

  correct_pred = tf.equal(tf.argmax(pred, 1),tf.argmax(y_
                                              vals, 1))
  accuracy = tf.reduce_mean(tf.cast(correct_pred,
                                              tf.float32))

  print("correct_pred:",sess.run(correct_pred))
  print("accuracy:",sess.run(accuracy))
```

Listing 2.22 contains code that is very similar to Listing 2.21: the main difference is that Listing 2.21 contains integer values for x and y, whereas the NumPy arrays pred and y_vals in Listing 2.22 contain decimal values that are between 0 and 1. The output from launching the code in Listing 2.22 is here:

```
argmax(pred,1):    [5 0 2]
argmax(y_vals,1):  [5 0 2]
correct_pred: [ True   True   True]
accuracy: 1.0
```

THE TF.ONE_HOT() API (REVISITED)

Chapter 1 contains an example of using the TF `one_hot()` API with TF arrays of numbers, and this section describes how to create a "one hot" encoding for categorical (i.e., non-numeric) data.

A feature that contains non-numeric values is called categorical or nominal data. Since neural networks expect numeric values as their input values, we must "map" non-numeric feature values into a corresponding set of numeric values. The term one-hot encoding involves the conversion of each non-numeric value into a vector that contains a single 1 (and zeroes elsewhere).

For example, suppose that we have a color variable the values of which are red, green, or blue. The one-hot encoding of this color variable happens to look like a 3x3 identity matrix, as shown here:

```
red,    green,   blue
1,      0,       0
0,      1,       0
0,      0,       1
```

Now, suppose that you have a dataset with six rows of data the color values of which are red, green, blue, red, green, and blue. Then the six rows would contain the following values (let's ignore the values of the other elements of these six rows):

```
1,      0,       0
0,      1,       0
0,      0,       1
1,      0,       0
0,      1,       0
0,      0,       1
```

INITIALIZING TF ARRAYS WITH RANDOM VALUES

Listing 2.23 displays the contents of `tf-rand-array.py`, which illustrates how to initialize TF variables with random values. This code sample is very useful for TensorFlow code that defines the contents of a hidden layer in a neural network.

LISTING 2.23: tf-rand-array.py

```
import tensorflow as tf
import numpy as np

W = tf.get_variable("W", [784, 256], initializer=tf.
truncated_normal_initializer(stddev=np.sqrt(2.0 / 784)))

Z = tf.get_variable("z", [4, 5], initializer=tf.random_
uniform_initializer(-1, 1))

init = tf.global_variables_initializer()

with tf.Session() as session:
  session.run(init)

  print("Z:",Z)
  print("W:",W)
  print("Z:",Z.eval())
  print("W:",W.eval())
```

Listing 2.23 initializes the variable W that has shape (784,256), the initial values of which are based on the tf.truncated_normal_initializer() API. The next portion of Listing 2.23 initializes the TF variable Z, this time with the tf.random_uniform_initializer() API. The next portion of Listing 2.23 is a with code block that initializes the TF variables and then prints the metadata for Z and W, followed by the actual numeric contents of Z and W. The output from launching the code in Listing 2.23 is here:

```
Z: <tf.Variable 'z:0' shape=(4, 5) dtype=float32_ref>
W: <tf.Variable 'W:0' shape=(784, 256) dtype=float32_ref>
Z: [[ 0.62888     0.07410955  0.42175293  0.9334245
                                            0.05753112]
 [ 0.85674214 -0.14664006 -0.90392494 -0.17587161
                                            0.47735047]
 [-0.6012428  -0.7233474   0.17535758  0.21230721
                                           -0.32793975]
 [-0.7565453   0.13265014 -0.87297916  0.52374625
                                            0.2544427 ]]
W: [[-0.00684283 -0.02124431  0.08983029 ...  0.04756822
                                           -0.05100965
    0.06723917]
 [-0.06147837  0.0054979   0.01208646 ...  0.03024297
                                           -0.06468915
   -0.01149644]
 [-0.01923586 -0.07037553  0.00995891 ...  0.08423746
                                           -0.02759983
   -0.00015794]
 ...
 [-0.01406402  0.00985579  0.06272404 ...  0.0116624
                                            0.00611281
   -0.01078485]
 [-0.05580244 -0.01177411  0.01413527 ...  0.04812782
                                           -0.02273556
```

```
  -0.08498681]
 [-0.02799775 -0.00868399 -0.03156329 ...  0.02155628
                                             -0.05653304
  -0.04880346]]
```

OTHER USEFUL TF APIS

In addition to the TF APIs that you have seen in this chapter, you will also encounter the following APIs, the names of which are intuitive. This section contains short code blocks that illustrate the syntax for these APIs, and you can find more detailed information in the online documentation.

The `tf.zeros()` API initializes a tensor with all zeroes, as shown here:

```
import tensorflow as tf
zeroes = tf.zeros([2, 3])
with tf.Session() as sess:
  print(sess.run(zeroes))
```

The output is here:

```
[[0. 0. 0.]
 [0. 0. 0.]]
```

The `tf.ones()` API initialize a tensor with all ones, as shown here:

```
import tensorflow as tf
ones = tf.ones ([2, 3])
with tf.Session() as sess:
  print(sess.run(ones))
```

The output is here:

```
[[1. 1. 1.]
 [1. 1. 1.]]
```

The `tf.fill()` API initializes a tensor with a specified numeric or string value, as shown here:

```
import tensorflow as tf
nines = tf.fill(dims=[2, 3], value=9)
pizza = tf.fill(dims=[2, 3], value="pizza")
with tf.Session() as sess:
  print(sess.run(nines))
  print(sess.run(pizza))
```

The output is here:

```
[[9 9 9]
 [9 9 9]]
[[b'pizza' b'pizza' b'pizza']
 [b'pizza' b'pizza' b'pizza']]
```

The `tf.unique()` API finds the unique numbers or strings (duplicate values are ignored) in a TF tensor, as shown here:

```
import tensorflow as tf

x = tf.constant([1, 1, 2, 4, 4, 4, 7, 8, 8])
val, idx = tf.unique(x)
y = tf.constant(['a','a','b','b','c','c'])
val2, idx2 = tf.unique(y)

with tf.Session() as sess:
  print("val: ",sess.run(val))
  print("idx: ",sess.run(idx))
  print("val2:",sess.run(val2))
  print("idx2:",sess.run(idx2))
```

The output is here:

```
val:  [1 2 4 7 8]
idx:  [0 0 1 2 2 2 3 4 4]
val2: [b'a' b'b' b'c']
idx2: [0 0 1 1 2 2]
```

The `tf.where()` API determines the location of a matching number (if any). For example, the following code block finds the location of the numbers 3 and 5 in the variable `t1`:

```
import tensorflow as tf

t1 = tf.constant([[1, 2, 3], [4, 5, 6]])
t2 = tf.where(tf.equal(t1, 3))
t3 = tf.where(tf.equal(t1, 5))

with tf.Session() as sess:
  print("t1:",sess.run(t1))
  print("t2:",sess.run(t2))
  print("t3:",sess.run(t3))
```

The output from launching the preceding code block is here:

```
t1: [[1 2 3]
 [4 5 6]]
t2: [[0 2]]
t3: [[1 1]]
```

Notice that `t1` has dimensions 2x3, and the number 3 appears in position 3 (which has index 2) of the first element (which has index 0). Hence, the result is an element that contains the one-dimensional array [0 2]. Similarly, the number 5 appears in `t1` in position 2 (which has index 1) of the second element (which has index 1). Hence, the result is an element that contains the one-dimensional array [1 1].

This concludes the portion of the chapter dedicated to frequently used TensorFlow APIs. The remaining portion of the chapter discusses TensorBoard, which is a web-based visualization tool (included in the TensorFlow distribution) that provides highly useful information for a TensorFlow graph.

USEFUL TENSORBOARD APIS

In Chapter 1, you learned that TensorBoard is a data visualization tool that is part of the TensorFlow distribution. TensorBoard also provides a "writer" for saving the contents of a TensorFlow graph to a file in a directory (that is specified by you). In addition, TensorBoard provides various APIs in order to insert the values of variables in a Tensorboard visualization.

After invoking a Python script with TensorBoard APIs, navigate to the directory that contains the graph-related files, and then launch TensorBoard from the command line, where the log directory is assumed to be /tmp/tf_logs:

```
tensorboard --logdir=/tmp/tf_logs
```

Next, launch a Chrome browser and navigate to this URL:

```
localhost:6006
```

When you saw the TensorFlow graph rendered in your browser, use your mouse to resize the graph, and double-click on nodes to "drill down" and find more information about each node.

The following subsections discuss some TensorBoard APIs, followed by code samples that illustrate how to save TensorFlow graphs in order to view them in TensorBoard.

The tf.name_scope() API

The concept of name scope is similar to namespaces in other languages (ex: Java and C#), and it's useful for two variables with same name but in different operations. In essence, specifying a name scope enables you to "group" variables together and display them in a "block" in TensorBoard. For example, the following code snippet defines b1 in the name scope called "scope_b1":

```
with tf.name_scope("scope_b1"):
  b1 = tf.add_n([a4, a5], name="b1")
```

The tf.summary_scalar() API

The tf.summary_scalar() API generates a tensor with a scalar value. Invoke this API in TensorFlow code, which will update the TensorFlow graph and the result will be visible in TensorBoard. Here are some examples:

```
tf.summary.scalar('stddev', stddev)
tf.summary.scalar('max', tf.reduce_max(var))
tf.summary.scalar('min', tf.reduce_min(var))
```

A BASIC TENSORBOARD EXAMPLE (1)

Listing 2.24 displays the contents of `tf-graph1.py`, which illustrates how to save a graph to a file and then display its contents in TensorBoard.

LISTING 2.24: tf-graph1.py

```
import tensorflow as tf

a = tf.add(1, 2,)
b = tf.multiply(a, 3)
c = tf.add(4, 5,)
d = tf.multiply(c, 6,)
e = tf.multiply(4, 5,)
f = tf.div(c, 6,)
g = tf.add(b, d)
h = tf.multiply(g, f)

with tf.Session() as sess:
    writer = tf.summary.FileWriter("output", sess.graph)
    print(sess.run(h))
    writer.close()
```

Listing 2.24 defines the variables a through h using arithmetic TensorFlow APIs (add, multiply, and div) in conjunction with integer values. Notice that the `with` code block saves the TensorFlow graph, along with the value of h (which is 63). The values of a through g are not saved in the TensorFlow graph. The output from launching the code in Listing 2.24 is here:

```
63
```

A BASIC TENSORBOARD EXAMPLE (2)

Listing 2.25 displays the contents of `tf-graph2.py`, which illustrates how to save a graph to a file and then display its contents in TensorBoard.

LISTING 2.25: tf-graph2.py

```
import tensorflow as tf

a = tf.add(1, 2, name="Add_these_numbers")
b = tf.multiply(a, 3)
c = tf.add(4, 5, name="And_These_ones")
d = tf.multiply(c, 6, name="Multiply_these_numbers")
e = tf.multiply(4, 5, name="B_add")
f = tf.div(c, 6, name="B_mul")
g = tf.add(b, d)
h = tf.multiply(g, f)

with tf.Session() as sess:
```

```
writer = tf.summary.FileWriter("output", sess.graph)
print(sess.run(h))
writer.close()
```

Listing 2.25 extends the code sample in Listing 2.24 by assigning a name to a, c, d, e, and f. The output from launching the code in Listing 2.25 is here:

63

A BASIC TENSORBOARD EXAMPLE (3)

Listing 2.26 displays the contents of `tf-graph3.py`, which illustrates how to save a graph to a file and then display its contents in TensorBoard.

LISTING 2.26: tf-graph3.py

```
import tensorflow as tf

with tf.name_scope("MyOperationGroup"):
    with tf.name_scope("Scope_A"):
        a = tf.add(1, 2, name="Add_these_numbers")
        b = tf.multiply(a, 3)
    with tf.name_scope("Scope_B"):
        c = tf.add(4, 5, name="And_These_ones")
        d = tf.multiply(c, 6, name="Multiply_these_numbers")

with tf.name_scope("Scope_C"):
    e = tf.multiply(4, 5, name="B_add")
    f = tf.div(c, 6, name="B_mul")
    g = tf.add(b, d)
    h = tf.multiply(g, f)

with tf.Session() as sess:
    writer = tf.summary.FileWriter("output", sess.graph)
    print(sess.run(h))
    writer.close()
```

Listing 2.26 extends the code sample in Listing 2.25 by specifying name scopes, including nested name scopes, for the variables a through h. The output from launching the code in Listing 2.26 is here:

63

If you want to view the graph in TensorBoard, launch the following command from the directory that contains Listing 2.27:

```
tensorboard -logdir=output
```

Launch a browser session and navigate to this URL: `localhost:6006`.

Now move your mouse pointer along the graph and you will see different values highlighted that are associated with the Gaussian curve that's located at the same place as your mouse location.

This concludes the introduction to TensorBoard, and performing an Internet search if you want to learn more about TensorBoard.

SUMMARY

In this chapter, you learned about some TensorFlow features, such as eager execution, which has a more Python-like syntax than "regular" TensorFlow syntax, tensor operations (such as multiplying tensors), and also how to create for loops and while loops in TensorFlow.

Next, you saw how to use the TensorFlow `tf.random_normal()` API for generating random numbers (which is useful for initializing the weights of edges in neural networks), followed by the `tf.argmax()` API for finding the index of each row (or column) that contains the maximum value in each row (or column), which is used for calculating the accuracy of the training process involving various algorithms. You also saw the `tf.range()` API, which is similar to the `NumPy np.linspace()` API.

In addition, you learned about the TF `reduce_mean()` and `equal()` APIs, both of which are involved in calculating the accuracy of a the training of a neural network (in conjunction with `tf.argmax()`). Next, you saw the TensorFlow `truncated_normal()` API, which is a variant of the `tf.random_normal()` API, and the TF `one_hot()` API for encoding data in a particular fashion. Moreover, you learned about the TensorFlow `reshape()` API, which you will see in any TensorFlow code that involves training a CNN.

In the second half of this chapter, you were introduced to TensorBoard, which is a very powerful visualization tool that is part of the TensorFlow distribution. You saw some code samples that invoke TensorBoard APIs alongside other TensorFlow APIs in order to augment the TensorFlow graph with supplemental information that was rendered in TensorBoard in a Web browser.

## CHAPTER 3

TENSORFLOW DATASETS

T his chapter contains various code samples that illustrate how to use TensorFlow `Datasets`, which support a rich set of operators that can simplify your TensorFlow code and enable you to process very large datasets (i.e., datasets that are too large to fit in memory). You will learn about operators (such as `filter()` and `map()`) that you can specify via "method chaining" as part of the definition of a TF dataset. In addition, you'll learn about TF `estimators` (in the `tf.estimator` namespace), TF `layers` (in the `tf.layers` namespace), and `TFRecords`.

Familiarity with lambda expressions (discussed later) and Reactive Programming will be very helpful for this chapter. In fact, this chapter will be very straightforward if you already have experience with `Observables` in `RxJS`, `RxAndroid`, `RxJava`, or some other environment that involves lazy execution.

The first part of this chapter briefly introduces you to TF `Datasets` and lambda expressions, along with some simple code samples. This section also introduces you to several types of `iterators` in TensorFlow, including "one shot" iterators and "reiterable" iterators.

The second part of this chapter discusses `TextLineDatasets`, which are very convenient for working with text files. The code samples in this section also use two of the four types of iterators that are discussed in the previous section.

The third part of this chapter discusses so-called "intermediate" or "lazy" operators, such as `filter()`, `flatmap()`, and `map()` operators. You can use *method chaining* in order to combine these operators, resulting in powerful code combinations that can significantly reduce the complexity of your TensorFlow code.

The final portion of the chapter briefly discusses TF `estimators`, which are implementations of various machine learning algorithms, as well as TF `layers` that provide an assortment of classes for Dense Neural Networks (DNN) and (Convolutional Neural Networks (CNN).

TENSORFLOW DATASETS

TensorFlow `Datasets` (accessible via the namespace `tf.data.Dataset`) are well-suited for creating asynchronous and optimized data pipelines. In brief, the TF `Dataset` API loads data from the disk (both images and text), applies optimized transformations, creates batches, and sends the batches to the GPU. Hence, the TF `Dataset` API is good for better GPU utilization. In addition, use `tf.functions` in TF 2.0 to utilize dataset asynchronous prefetching/streaming features fully.

A TF `Dataset` acts as a container-like "wrapper" around a dataset, somewhat analogous to a `Pandas DataFrame`. A very large dataset can therefore involve a very large TF `Dataset`. A TF `Dataset` can also represent an input pipeline as a collection of elements (i.e., a nested structure of tensors), along with a "logical plan" of transformations that act on those elements. For example, you can define a TF `Dataset` that initially contains the lines of text in a text file, then extract the lines of text that start with a "#" character, and then display only the first three lines. Creating this pipeline is easy: create a TF `Dataset` and then chain the lazy operators `filter()` and `take()`, which is similar to an example that you will see later in this chapter.

Basic Steps for TensorFlow `Datasets`

Perform the following three steps in order to create and process the contents of a TF `Dataset`:

1. Import Data
2. Create an Iterator
3. Consume the Data

There are many ways to populate a TF `Dataset` with data that can be retrieved from multiple sources. For simplicity, the code samples in the first part of this chapter perform the following steps: first create a TF `Dataset` instance with an initialized `NumPy` array of data; second, define an iterator instance in order to iterate through the TF `Dataset`; and third, use the iterator to access the elements of the dataset (and, in some cases, supply those elements to a TF model).

A Simple TensorFlow Dataset

Listing 3.1 displays the contents of `tf-numpy-dataset.py`, which illustrate how to create a TF `Dataset` from a `NumPy` array of numbers. Although this code sample is minimalistic, it's the starting point in various other code samples in this chapter.

LISTING 3.1: tf-numpy-dataset.py

```
import tensorflow as tf
import numpy as np

x = np.arange(0, 10)

# make a dataset from a numpy array
dataset = tf.data.Dataset.from_tensor_slices(x)
```

Listing 3.1 contains two `import` statements and then initializes the variable x as a `NumPy` array with the integers from 0 through 9 inclusive. The variable `dataset` is initialized as a TF `Dataset` that's based on the contents of the variable x.

Note that nothing else happens in Listing 3.1: later you will see more meaningful code samples involving TF `Datasets`.

TENSORFLOW RAGGED CONSTANTS AND TENSORS (OPTIONAL)

The section is marked "optional" because the code samples require TF 2. If you launch the code samples using TF 1.x, you will see the following error message:

```
AttributeError: 'module' object has no attribute
                                     'RaggedTensor'
```

As you probably know, every element in a multi-dimensional array has the same dimensions. For example, a 2x3 array contains two rows and three columns: each row is a 1x3 vector, and each column is a 2x1 vector. As another example, a 2x3x4 array contains two 3x4 arrays (and the same logic applies to each 3x4 array).

On the other hand, a *ragged constant* is a set of elements that has different lengths. You can think of ragged constants as a generalization of "regular" datasets.

Listing 3.2 displays the contents of `tf-raggedtensors1.py`, which illustrates how to define a ragged dataset and then iterate through its contents.

LISTING 3.2: tf-raggedtensors1.py

```
import tensorflow as tf

digits = tf.ragged.constant([[3, 1, 4, 1], [], [5, 9, 2],
                                                [6], []])
words = tf.ragged.constant([["Bye", "now"], ["thank",
                                "you", "again", "sir"]])

print(tf.add(digits, 3))
print(tf.reduce_mean(digits, axis=1))
print(tf.concat([digits, [[5, 3]]], axis=0))
print(tf.tile(digits, [1, 2]))
print(tf.strings.substr(words, 0, 2))
```

Listing 3.2 defines two ragged constants `digits` and `words` consisting of integers and strings, respectively. The remaining portion of Listing 3.2 is five `print()` statements that applies various operations to these two datasets and then displays the results.

The first `print()` statement adds the value 3 to every number in the `digits` dataset, and the second `print()` statement computes the row-wise average of the elements of the `digits` dataset because `axis=1` (whereas `axis=0` performs column-wise operations).

The third `print()` statement appends the element `[[5,3]]` to the `digits` dataset, and performs this operation in a column-wise fashion (because `axis=0`). The fourth `print()` statement "doubles" each non-empty element of the `digits` dataset. Finally, the fifth `print()` statement extracts the first two characters from each string in the `words` dataset.

The output from launching the code in Listing 3.2 is here:

```
<tf.RaggedTensor [[6, 4, 7, 4], [], [8, 12, 5], [9], []]>
tf.Tensor([2.25            nan 5.33333333 6.
                 nan], shape=(5,), dtype=float64)
<tf.RaggedTensor [[3, 1, 4, 1], [], [5, 9, 2], [6], [], [5,
                                                          3]]>
<tf.RaggedTensor [[3, 1, 4, 1, 3, 1, 4, 1], [], [5, 9, 2,
                                  5, 9, 2], [6, 6], []]>
<tf.RaggedTensor [[b'By', b'no'], [b'th', b'yo', b'ag',
                                               b'si']]>
```

Listing 3.3 displays the contents of `tf-ragged-tensors.py` that illustrates how to define a ragged tensor in TensorFlow.

LISTING 3.3: tf-ragged-tensors.py

```python
import tensorflow as tf

x1 = tf.RaggedTensor.from_row_splits(
        values=[1, 2, 3, 4, 5, 6, 7, 8, 9, 10],
        row_splits=[0, 5, 10])
print("x1:",x1)

x2 = tf.RaggedTensor.from_row_splits(
        values=[1, 2, 3, 4, 5, 6, 7, 8, 9, 10],
        row_splits=[0, 4, 7, 10])
print("x2:",x2)

x3 = tf.RaggedTensor.from_row_splits(
        values=[1, 2, 3, 4, 5, 6, 7, 8],
        row_splits=[0, 4, 4, 7, 8, 8])
print("x3:",x3)
```

Listing 3.3 defines the TF ragged tensors `x1`, `x2`, and `x3` that are based on the integers from 0 to 10 inclusive. The `values` parameter specifies a set of values that will be "split" into a set of vectors, using the numbers in the `row_splits` parameter for the start index and the end index of each vector.

For example, x1 specifies `row_split` with the value [0,5,10], which determines two vectors: the vector whose values are from index 0 through index 4 of x1, and the vector whose values are from index 5 through index 9 of x1. The contents of those two vectors are [1, 2, 3, 4, 5] and [6, 7, 8, 9, 10], respectively (see the output below).

As another example, x2 specifies `row_splits` with the value [0,4,7,10], which determines three vectors: the vector whose values are from index 0 through index 3 of x1, the vector whose values are from index 4 through index 6 of x1, and the vector whose values are from index 7 through index 9 of x1. The contents of those two vectors are [1,2,3,4], [5,6,7], and [8, 9, 10], respectively (see the output below).

You can perform a similar analysis for x3, keeping mind that the vector for which the start index and end index are [4,4] is an empty vector. The output from launching the code in Listing 3.3 is here:

```
x1: <tf.RaggedTensor [[1, 2, 3, 4, 5], [6, 7, 8, 9, 10]]>
x2: <tf.RaggedTensor [[1, 2, 3, 4], [5, 6, 7], [8, 9, 10]]>
x3: <tf.RaggedTensor [[1, 2, 3, 4], [], [5, 6, 7], [8], []]>
```

If you want to generate a list of values, invoke the `to_list()` operator. For instance, suppose you define x4 as follows:

```
x4 = tf.RaggedTensor.from_row_splits(
        values=[1, 2, 3, 4, 5, 6, 7, 8, 9, 10],
        row_splits=[0, 5, 10]).to_list()
print("x4:",x4)
```

The output from the preceding code snippet is here (which you can compare with the output for x1 in the preceding output block):

```
x4: [[1, 2, 3, 4, 5], [6, 7, 8, 9, 10]]
```

You can also create higher-dimensional ragged tensors in TF. For example, the following code snippet creates a two-dimensional ragged tensor in TF:

```
RaggedTensor.from_nested_row_splits(
        flat_values=[3,1,4,1,5,9,2,6],
        nested_row_splits=([0,3,3,5], [0,4,4,7,8,8])).to_list()
```

The preceding code snippet generates the following output:

```
[[[3, 1, 4, 1], [], [5, 9, 2]], [], [[6], []]]
```

WHAT ARE LAMBDA EXPRESSIONS?

In brief, a *lambda expression* is an anonymous function. Use lambda expressions to define local functions that can be passed as arguments or returned as the value of function calls.

Informally, a lambda expression takes an input variable and performs some type of operation (specified by you) on that variable. For example, here's a "bare bones" lambda expression that adds the number 1 to an input variable x:

```
lambda x: x + 1
```

You can use the preceding lambda expression in a valid TensorFlow code snippet, as shown here (dx is a TensorFlow Dataset that is defined elsewhere):

```
dx.map(lambda x: x + 1)
```

Even if you are unfamiliar with TensorFlow Datasets or the map() operator, you can still understand the preceding code snippet.

The next section contains a complete TensorFlow code sample that illustrates how to use the preceding lambda expression.

A LAMBDA EXPRESSION IN TENSORFLOW

Listing 3.4 displays the contents of tf-plusone.py, which illustrates how to use a lambda expression to add the number 1 to the elements of a TF Dataset.

LISTING 3.4: tf-plusone.py

```
import tensorflow as tf
import numpy as np

x = np.arange(0, 10)

# create dataset object from Numpy array
dx = tf.data.Dataset.from_tensor_slices(x)

dx.map(lambda x: x + 1)

# create a one-shot iterator
iterator = dx.make_one_shot_iterator()

# extract an element
next_element = iterator.get_next()

with tf.Session() as sess:
  for i in range(10):
    val = sess.run(next_element)
    print("val:",val)
```

Listing 3.4 initializes the variable x as NumPy array consisting of the integers from 0 through 9 inclusive. Next, the variable dx is initialized as a TF Dataset that is created from the contents of the variable x. Notice how dx.map() then defines a lambda expression that adds one to each input value, which consists of the integers from 0 to 9 in this example.

The next code snippet defines the variable iterator as a "one shot" iterator, followed by the variable next_element that is populated with the first integer in dx (which is the integer 0).

The final portion of Listing 3.4 iterates through the element of dx and displays their values. The output from launching the code in Listing 3.4 is here:

```
('val:', 0)
('val:', 1)
('val:', 2)
('val:', 3)
('val:', 4)
('val:', 5)
('val:', 6)
('val:', 7)
('val:', 8)
('val:', 9)
```

WHAT ARE ITERATORS?

An *iterator* is similar to a "cursor" in other languages. You can think of an iterator as something that "points" to an item in a dataset. By way of analogy, if you have a linked list of items, an iterator is analogous to a pointer that "points" to the first element in the list, and each time you move the pointer to the next item in the list, you are "advancing" the iterator.

Working with datasets and iterators involves the steps:

1. create a dataset
2. create an iterator (see next section)
3. "point" the iterator to the dataset
4. print the contents of the current item
5. "advance" the iterator to the next item
6. go to Step 4 if there are more items

Notice that Step 6 specifies "if there are more items," which you can handle via a try/except block (shown later in this chapter) when the iterator goes beyond the last item in the dataset. This technique is very useful because it obviates the need to know the number of items in a dataset. TensorFlow provides several types of iterators, as discussed in the next section.

TensorFlow Iterators

TensorFlow supports four types of iterators, as listed here:

1. One shot
2. Initializable
3. Reinitializable
4. Feedable

A *one-shot iterator* can iterate only once through a dataset. After we reach the end of the dataset, the iterator will no longer yield elements; instead, it will raise an Exception. For example, if dx is an instance of tf.data.Dataset, then the following code snippet defines a one-shot iterator:

```
iterator = dx.make_one_shot_iterator()
```

An *initializable iterator* can be dynamically updated: invoke its initializer operation and pass new data via the parameter feed_dict. If dx is an instance of tf.data.Dataset, then the following code snippet defines a reusable iterator:

```
iterator = dx.make_initializable_iterator()
```

A *reinitializable iterator* can be initialized from a different Dataset. This type of iterator is very useful for training datasets that require some additional transformation, such as shuffling its contents.

A *feedable iterator* allows you to select from different iterators: this type of iterator is essentially a "selector" to select an iterator from a collection of iterators.

The code samples in this chapter involve either one-shot iterators or Initializable iterators. The next section contains a code sample that illustrates how to define a TF Dataset and a reusable iterator in TensorFlow.

TENSORFLOW REUSABLE ITERATORS

Listing 3.5 displays the contents of tf-init-iterator.py, which illustrates how to define a reusable iterator in TensorFlow.

LISTING 3.5: tf-init-iterator.py

```
import tensorflow as tf
import numpy as np

x = np.arange(0, 10)
dx = tf.data.Dataset.from_tensor_slices(x)

# create an initializable iterator
iterator = dx.make_initializable_iterator()

# extract an element
next_element = iterator.get_next()

with tf.Session() as sess:
  sess.run(iterator.initializer)
  for i in range(15):
    val = sess.run(next_element)
    print(val)
    if i % 9 == 0 and i > 0:
      sess.run(iterator.initializer)
```

Listing 3.5 initializes the variable x as NumPy array consisting of the integers from 0 through 9 inclusive. Next, the variable dx is initialized as a TF Dataset that is created from the contents of the variable x. The next code snippet defines the variable iterator as an "initializable" iterator, followed by the variable `next_element` that is populated with the first integer in dx (which is the integer 0).

The final portion of Listing 3.5 involves a `for` loop that iterates through the elements of `dx` and displays their values. Notice that the loop iterates through 15 values, whereas dx contains only 10 elements. What will happen?

The answer lies in the conditional logic in the `for` loop: when the variable i equals 9, the iterator variable is "reset" to its initial value, thereby preventing an error. You can test this situation by "commenting out" the `if` statement, and when you launch the code you will see the following error message:

```
OutOfRangeError (see above for traceback): End of sequence
[[node IteratorGetNext (defined at tf-init-iterator.py:11) ]]
```

Reset the code in Listing 3.5 to its original contents and you will see the following output when you launch the code in Listing 3.5 is here:

```
0
1
2
3
4
5
6
7
8
9
0
1
2
3
4
```

THE TENSORFLOW FILTER() OPERATOR

Listing 3.6 displays the contents of `tf-filter.py`, which illustrates how to use the filter`()` operator in TensorFlow with a "one shot" iterator.

LISTING 3.6: tf-filter.py

```
import tensorflow as tf
import numpy as np

#def filter_fn(x):
#    return tf.reshape(tf.not_equal(x % 2, 1), [])

x = np.array([1,2,3,4,5,6,7,8,9,10])

ds = tf.data.Dataset.from_tensor_slices(x)
ds = ds.filter(lambda x: tf.reshape(tf.not_equal(x%2,1), []))
```

```
#ds = ds.filter(filter_fn)

iterator = ds.make_one_shot_iterator()
next_element = iterator.get_next()

with tf.Session() as sess:
  try:
    while True:
      value = sess.run(next_element)
      print("value:",value)
  except tf.errors.OutOfRangeError:
    pass
      sess.run(iterator.initializer)
```

Listing 3.6 initializes the variable x as a NumPy array consisting of the integers from 1 through 10 inclusive. Next, the variable ds is initialized as a TF Dataset that is created from the contents of the variable x. The next code snippet invokes the filter() operator, inside of which a lambda expression returns even numbers because of this expression:

```
tf.not_equal(x%2,1)
```

The next code snippet defines the variable iterator as a "one shot" iterator, followed by the variable next_element that is populated with the first integer in dx (which is the integer 0).

The final portion of Listing 3.6 contains a try/except block, inside of which there is a while True code block that iterates through the elements of the dataset until a tf.errors.OutOfRangeError error is reached. When this error occurs, the code exits the loop gracefully. The output from launching the code in Listing 3.6 is here:

```
('value:', 2)
('value:', 4)
('value:', 6)
('value:', 8)
('value:', 10)
```

TensorFlow also supports a TextLineDataset that is useful for processing the contents of text files. The next several sections contain code samples that show you how to define a TextLineDataset and perform simple operations.

TENSORFLOW TEXTLINEDATASET (1)

Listing 3.7 displays the contents of file.txt, which is referenced in the code in Listing 3.8.

LISTING 3.7: file.txt

```
this is file line #1
this is file line #2
this is file line #3
```

```
this is file line #4
this is file line #5
```

Listing 3.8 displays the contents of tf-textlinedataset1.py, which illustrates how to define a TextLineDataset in TensorFlow.

LISTING 3.8: tf-textlinedataset1.py

```
import tensorflow as tf
import numpy as np

dataset = tf.data.TextLineDataset("file.txt")
dataset = dataset.map(lambda string: tf.string_
                                    split([string]).values)

iterator = dataset.make_initializable_iterator()
next_element = iterator.get_next()
init_op = iterator.initializer

with tf.Session() as sess:
  sess.run(init_op)
  print(sess.run(next_element))
```

Listing 3.8 initializes the variable dataset as a TF Dataset that contains the contents of the text file file.txt. The next code snippet defines the variable iterator as an "initializable" iterator, followed by the variable next_element that is populated with the first line of text in dataset.

The final portion of Listing 3.8 prints one "tokenized" line of text from the file file.txt, as shown here:

```
['this' 'is' 'file' 'line' '#1']
```

TENSORFLOW TEXTLINEDATASET (2)

Listing 3.9 displays the contents of tf-textlinedataset2.py, which illustrates another way to define a TextLineDataset in TensorFlow.

LISTING 3.9: tf-textlinedataset2.py

```
import tensorflow as tf
import numpy as np

dataset = tf.data.TextLineDataset("file.txt")
dataset = dataset.map(lambda string: tf.string_
                                    split([string]).values)

iterator = dataset.make_one_shot_iterator()
next_element = iterator.get_next()

with tf.Session() as sess:
  print(sess.run(next_element))
  print(sess.run(next_element))
  print(sess.run(next_element))
```

Listing 3.9 initializes the variable dataset as a TF Dataset that contains the contents of the text file file.txt. The next code snippet defines a lambda expression that tokenizes each line of text in the file file.txt.

The next code snippet defines the variable iterator as an "initializable" iterator, followed by the variable next_element that is populated with the first line of text in the file file.txt.

The final portion of Listing 3.9 prints three "tokenized" lines of text from the file file.txt, as shown here:

```
['this' 'is' 'file' 'line' '#1']
['this' 'is' 'file' 'line' '#2']
['this' 'is' 'file' 'line' '#3']
```

TENSORFLOW TEXTLINEDATASET (3)

Listing 3.10 displays the contents of tf-textlinedataset3.py, which illustrates a third way to define a TextLineDataset in TensorFlow.

LISTING 3.10: tf-textlinedataset3.py

```
import tensorflow as tf
import itertools

dataset = tf.data.TextLineDataset("file.txt")
dataset = dataset.map(lambda string: tf.string_
                                  split([string]).values)

iterator = dataset.make_one_shot_iterator()
next_element = iterator.get_next()

with tf.Session() as sess:
  for i in range(5):
    print(sess.run(next_element))
```

Listing 3.10 initializes the variable dataset as a TF Dataset that contains the contents of the text file file.txt. The next code snippet defines a lambda expression that tokenizes each line of text in the file file.txt.

The next code snippet defines the variable iterator as a "one shot" iterator, followed by the variable next_element that is populated with the first line of text in the dataset variable.

The final portion of Listing 3.10 contains a for loop that prints five "tokenized" lines of text from the file file.txt, as shown here:

```
['this' 'is' 'file' 'line' '#1']
['this' 'is' 'file' 'line' '#2']
['this' 'is' 'file' 'line' '#3']
['this' 'is' 'file' 'line' '#4']
['this' 'is' 'file' 'line' '#5']
```

The companion disc also contains the files tf-textlinedataset4.py and tf-textlinedataset4.py that use tf.data.TextLineDataset and lazy operators.

THE TENSORFLOW BATCH() OPERATOR (1)

Listing 3.11 displays the contents of `tf-batch1.py`, which illustrates how to use the `batch()` operator in TensorFlow with a reusable iterator.

LISTING 3.11: tf-batch1.py

```
import tensorflow as tf
import numpy as np

x = np.arange(0, 33)
dx = tf.data.Dataset.from_tensor_slices(x).batch(3)
iterator = dx.make_initializable_iterator()

with tf.Session() as sess:
  sess.run(iterator.initializer)
  for i in range(15):
    next_element = iterator.get_next()
    val = sess.run(next_element)
    print(val)
    if (i + 1) % (10 // 3) == 0 and i > 0:
      sess.run(iterator.initializer)
```

Listing 3.11 initializes the variable x as NumPy array consisting of the integers from 0 through 32 inclusive. Next, the variable dx is initialized as a TF Dataset that is created from the contents of the variable x. Notice that the definition of x involves method chaining by "tacking on" the `batch(3)` operator as part of the definition of dx.

The next code snippet defines the variable iterator as an "initializable" iterator, followed by the variable next_element that is populated with the first integer in dx (which is the integer 0).

The final portion of Listing 3.11 contains a loop that executes 15 times, and prints five "blocks" of numbers, where a block consists of the following output:

```
[0 1 2]
[3 4 5]
[6 7 8]
```

The reason for this "chunked" effect is because of the conditional logic that "resets" the iterator to the first element of the dataset, as shown here:

```
if (i + 1) % (10 // 3) == 0 and i > 0:
  sess.run(iterator.initializer)
```

Now launch the code in Listing 3.11 to see the output in its entirety, as shown here:

```
[0 1 2]
[3 4 5]
[6 7 8]
[0 1 2]
[3 4 5]
[6 7 8]
```

```
[0 1 2]
[3 4 5]
[6 7 8]
[0 1 2]
[3 4 5]
[6 7 8]
[0 1 2]
[3 4 5]
[6 7 8]
```

THE TENSORFLOW BATCH() OPERATOR (2)

Listing 3.12 displays the contents of tf-batch2.py, which illustrates how to use the batch() operator in TensorFlow with a one-shot iterator.

LISTING 3.12: tf-batch2.py

```
import tensorflow as tf
import numpy as np

x = np.arange(0, 33)
dx = tf.data.Dataset.from_tensor_slices(x).batch(3)
iterator = dx.make_one_shot_iterator()
next_element = iterator.get_next()

with tf.Session() as sess:
  for i in range(11):
    val = sess.run(next_element)
    print(val)
```

Listing 3.12 initializes the variable x as NumPy array consisting of the integers from 0 through 32 inclusive. Next, the variable dx is initialized as a TF Dataset that is created from the contents of the variable x. Notice how method chaining is performed by "tacking on" the batch(3) operator as part of the definition of dx.

The next code snippet defines the variable iterator as a "one shot" iterator, followed by the variable next_element that is populated with the first integer in dx (which is the integer 0).

The final portion of Listing 3.12 contains a loop that executes 11 times, and during each iteration the loop prints eleven "triples" of consecutive integers, starting with the triple [0 1 2]. The output from launching the code in Listing 3.12 is here:

```
[0  1  2]
[0  1  2]
[3  4  5]
[6  7  8]
[ 9 10 11]
[12 13 14]
[15 16 17]
```

```
[18 19 20]
[21 22 23]
[24 25 26]
[27 28 29]
[30 31 32]
```

The companion disc contains `tf-batch3.py`, `tf-batch4.py`, and `tf-batch5.py` that illustrate variations of the preceding code sample. Experiment with the code by changing the hard-coded values and then see if you can correctly predict the output.

THE TENSORFLOW MAP() OPERATOR (1)

Listing 3.13 displays the contents of `tf-map.py`, which illustrate how to use the `map()` operator in TensorFlow with a one-shot iterator.

LISTING 3.13: tf-map.py

```
import tensorflow as tf
import numpy as np

# a simple Numpy array
x = np.array([[1],[2],[3],[4]])

# make a dataset from a numpy array
dataset = tf.data.Dataset.from_tensor_slices(x)

# a lambda expression to double each value
dataset = dataset.map(lambda x: x*2)

# define an iterator
iter = dataset.make_one_shot_iterator()
el = iter.get_next()

with tf.Session() as sess:
  for _ in range(len(x)):
    print("value:",sess.run(el))
```

Listing 3.13 initializes the variable x as a `NumPy` array consisting of four elements, where each element is a 1x1 array consisting of the numbers 1, 2, 3, and 4. Next, the variable `dataset` is initialized as a TF `Dataset` that is created from the contents of the variable x. Notice how `dataset.map()` then defines a lambda expression that doubles each input value, which consists of the integers from 1 to 4 in this example.

The next code snippet defines the variable `iter` as a "one shot" iterator, followed by the variable el that is populated with the first element in the variable dataset.

The final portion of Listing 3.13 iterates through the element of `dataset` and displays their values. The output from launching the code in Listing 3.13 is here:

```
('value:', array([2]))
('value:', array([4]))
```

```
('value:', array([6]))
('value:', array([8]))
```

THE TENSORFLOW MAP() OPERATOR (2)

Listing 3.14 displays the contents of tf-map2.py, which illustrates how to invoke the map() operator three times in TensorFlow with a one-shot iterator.

LISTING 3.14: tf-map2.py

```
import tensorflow as tf
import numpy as np

# a simple Numpy array
x = np.array([[1],[2],[3],[4]])

# make a dataset from a Numpy array
dataset = tf.data.Dataset.from_tensor_slices(x)

# METHOD #1: THE LONG WAY
# a lambda expression to double each value
#dataset = dataset.map(lambda x: x*2)
# a lambda expression to add one to each value
#dataset = dataset.map(lambda x: x+1)
# a lambda expression to cube each value
#dataset = dataset.map(lambda x: x**3)

# METHOD #2: A SHORTER WAY
dataset = dataset.map(lambda x: x*2).map(lambda x: x+1).
                                      map(lambda x: x**3)

# define an iterator
iter = dataset.make_one_shot_iterator()
el = iter.get_next()

with tf.Session() as sess:
  for _ in range(len(x)):
    print("value:",sess.run(el))
```

Listing 3.14 initializes the variable x as a NumPy array consisting of four elements, where each element is a 1x1 array consisting of the numbers 1, 2, 3, and 4. Next, the variable dataset is initialized as a TF Dataset that is created from the contents of the variable x.

The next portion of Listing 3.14 is a "commented out" code block that consists of three lambda expressions, followed by a code snippet (shown in bold) that is a more compact way of defining the same three lambda expressions:

```
dataset = dataset.map(lambda x: x*2).map(lambda x: x+1).
                                      map(lambda x: x**3)
```

The preceding code snippet transforms each input value by first doubling the value, then adding one to the first result, and then cubing the second result.

Although method chaining is a concise way to chain operators, invoking dozens of lazy operators in a single (very long) line of code can quickly become difficult to understand, whereas writing code using the "longer way" would be easier to debug.

A suggestion: start with each lazy operator in a separate line of code, and after you are satisfied that the individual results are correct, *then* use method chaining to combine the operators in a single line of code (up to a maximum of four or five lazy operators).

The next code snippet defines the variable `iter` as a "one shot" iterator, followed by the variable el that is populated with the first element in the variable dataset.

The final portion of Listing 3.16 contains a `for` loop that iterates through the transformed values and displays their values. The output from launching the code in Listing 3.14 is here:

```
('value:', array([27]))
('value:', array([125]))
('value:', array([343]))
('value:', array([729]))
```

THE TENSORFLOW FLATMAP() OPERATOR (1)

In addition to the TF `map()` operator, TF also supports the TF `flat_map()` operator. However, the TF `Dataset.map()` and TF `Dataset.flat_map()` operators expect functions with different signatures. Specifically, `Dataset.map()` takes a function that maps a single element of the input dataset to a single new element, whereas `Dataset.flat_map()` takes a function that maps a single element of the input dataset to a `Dataset` of elements.

Listing 3.15 displays the contents of `tf-flatmap1.py`, which illustrates how to use the `flatmap()` operator in TensorFlow with a one-shot iterator.

LISTING 3.15: tf-flatmap1.py

```
import tensorflow as tf
import numpy as np

x = np.array([[1,2,3], [4,5,6], [7,8,9]])

ds = tf.data.Dataset.from_tensor_slices(x)
ds.flat_map(lambda x: tf.data.Dataset.from_tensor_slices(x))

iterator = ds.make_one_shot_iterator()
next_element = iterator.get_next()

with tf.Session() as sess:
  for i in range(3):
    value = sess.run(next_element)
    print("value:",value)
    next_element = iterator.get_next()
```

Listing 3.15 initializes the variable x as a NumPy array consisting of three elements, where each element is a 1x3 array of numbers. Next, the variable ds is initialized as a TF Dataset that is created from the contents of the variable x.

The next code snippet defines the variable iterator as a "one shot" iterator, followed by the variable next_element that is populated with the first element in the variable ds.

The final portion of Listing 3.15 iterates through the element of dataset and displays their values. The output from launching the code in Listing 3.15 is here:

```
('value:', array([1, 2, 3]))
('value:', array([4, 5, 6]))
('value:', array([7, 8, 9]))
```

THE TENSORFLOW FLATMAP() OPERATOR (2)

The code in the previous section works fine, but there is one drawback: there is a hard-coded value 3 in the code block that displays the elements of the dataset. This section removes the hard-coded value.

Listing 3.16 displays the contents of tf-flatmap2.py, which illustrates how to use the flatmap() operator in TensorFlow with a one-shot iterator, and then iterate through the elements of the dataset.

LISTING 3.16: tf-flatmap2.py

```
import tensorflow as tf
import numpy as np

x = np.array([[1,2,3], [4,5,6], [7,8,9]])

ds = tf.data.Dataset.from_tensor_slices(x)
ds.flat_map(lambda x: tf.data.Dataset.from_tensor_slices(x))

iterator = ds.make_one_shot_iterator()
next_element = iterator.get_next()

with tf.Session() as sess:
  try:
    while True:
      value = sess.run(next_element)
      print("value:",value)
  except tf.errors.OutOfRangeError:
    pass
```

Listing 3.16 initializes the variable x as a NumPy array consisting of three elements, where each element is a 1x3 array of numbers. Next, the variable ds is initialized as a TF Dataset that is created from the contents of the variable x.

The next code snippet defines the variable iterator as a "one shot" iterator, followed by the variable next_element that is populated with the first element in the variable ds.

The final portion of Listing 3.18 iterates through the element of `dataset` and displays their values. Notice that this code block contains a `try/except` block, inside of which there is a `while True` code block that iterates through the elements of the dataset until a `tf.errors.OutOfRangeError` error is reached. When this error occurs, the code exits the loop gracefully. The output from launching the code in Listing 3.16 is the same as the output from Listing 3.15:

```
('value:', array([1, 2, 3]))
('value:', array([4, 5, 6]))
('value:', array([7, 8, 9]))
```

THE TENSORFLOW FLAT_MAP() AND FILTER() OPERATORS

Listing 3.17 displays the contents of `comments.txt` and Listing 3.18 displays the contents of `tf-flatmap-filter.py`, which illustrate how to use the `filter()` operator in TensorFlow with a one-shot iterator.

LISTING 3.17: comments.txt

```
#this is file line #1
#this is file line #2
this is file line #3
this is file line #4
#this is file line #5
```

LISTING 3.18: tf-flatmap-filter.py

```python
import tensorflow as tf
import numpy as np

filenames = ["comments.txt"]

dataset = tf.data.Dataset.from_tensor_slices(filenames)

# 1) Use Dataset.flat_map() to transform each file as
# a separate nested dataset, and then concatenate their
# contents sequentially into a single "flat" dataset.
# 2) Skip the first line (header row).
# 3) Filter out lines beginning with "#" (comments).

dataset = dataset.flat_map(
    lambda filename: (
        tf.data.TextLineDataset(filename)
        .skip(1)
        .filter(lambda line: tf.not_equal(tf.strings.
                                substr(line, 0, 1), "#"))))

iterator = dataset.make_one_shot_iterator()
next_element = iterator.get_next()

with tf.Session() as sess:
  print(sess.run(next_element))
  print(sess.run(next_element))
```

Listing 3.18 defines the variable `filenames` as an array of text filenames, which in this case consists of just one file, called `comments.txt` (shown in Listing 3.17). Next, the variable `dataset` is initialized as a TF `Dataset` that contains the contents of `comments.txt`.

The next section of Listing 3.18 is a code block that explains the purpose of the next code block, which involves a moderately complex set of operations that are executed via method chaining in order to transform the contents of the variable dataset.

Specifically, the `dataset.flat_map()` operator "flattens" the contents of its input, which means that the input files (remember, it's just `comments.txt` in this example) are treated as an input stream. Second, the lambda expression "maps" each filename to an instance of the TF `tf.data.TextLineDataset` class. Third, the `skip(1)` operator skips the first line of each input file (`skip(n)` would skip the first n lines). Fourth, the `filter()` operator returns each input line (which is a line of text from each file) if and only if the input line does *not* start with the "#" character.

The next portion of Listing 3.18 defines the variable `iterator` as a "one shot" iterator, followed by the variable `next_element` that is populated with the first element in the variable `dataset`.

The final portion of Listing 3.18 prints two lines of output, which might seem anti-climatic after defining such a fancy set of transformations!

Launch the code in Listing 3.18 and you will see the following output:

```
this is file line #3
this is file line #4
```

THE TENSORFLOW REPEAT() OPERATOR

Listing 3.19 displays the contents of `tf-repeat.py`, which illustrates how to use the `repeat()` operator in TensorFlow.

LISTING 3.19: tf-repeat.py

```
import tensorflow as tf
import numpy as np

ds1 = tf.data.Dataset.from_tensor_slices(tf.range(4))
ds1 = ds1.repeat(2)

iterator = ds1.make_one_shot_iterator()
next_element = iterator.get_next()

with tf.Session() as sess:
  try:
    while True:
      value = sess.run(next_element)
      print("value:",value)
  except tf.errors.OutOfRangeError:
    pass
```

Listing 3.19 initializes the variable ds1 as a TF Dataset that is created from the integers between 0 and 3 inclusive. The next code snippet "tacks on" the repeat() operator to ds1, which has the effect of appending the contents of ds1 to itself.

The next code snippet defines the variable iterator as a "one shot" iterator, followed by the variable next_element that is populated with the first integer in ds1 (which is the integer 0).

The final portion of Listing 3.21 contains a try/except block, inside of which there is a while True code block that iterates through the elements of the dataset until an tf.errors.OutOfRangeError error is reached. When this error occurs, the code exits the loop gracefully. The output from launching the code in Listing 3.19 is here:

```
('value:', 0)
('value:', 1)
('value:', 2)
('value:', 3)
('value:', 0)
('value:', 1)
('value:', 2)
('value:', 3)
```

THE TENSORFLOW TAKE() OPERATOR

Listing 3.20 displays the contents of tf-take.py, which illustrates how to use the take() operator in TensorFlow.

LISTING 3.20: tf-take.py

```
import tensorflow as tf
import numpy as np

ds1 = tf.data.Dataset.from_tensor_slices(tf.range(8))
ds1 = ds1.take(5)

iterator = ds1.make_one_shot_iterator()
next_element = iterator.get_next()

with tf.Session() as sess:
  try:
    while True:
      value = sess.run(next_element)
      print("value:",value)
  except tf.errors.OutOfRangeError:
    pass
```

Listing 3.20 initializes the variable ds1 as a TF Dataset that is created from the integers between 0 and 7 inclusive. The next code snippet "tacks on" the take() operator to ds1, which has the effect of limiting the output to the first five integers.

The next code snippet defines the variable `iterator` as a "one shot" iterator, followed by the variable `next_element` that is populated with the first integer in `ds1` (which is the integer 0).

The final portion of Listing 3.20 contains a `try/except` block, inside of which there is a `while True` code block that iterates through the elements of the dataset until a `tf.errors.OutOfRangeError` error is reached. When this error occurs, the code exits the loop gracefully.

The output from launching the code in Listing 3.20 is here:

```
('value:', 0)
('value:', 1)
('value:', 2)
('value:', 3)
('value:', 4)
```

COMBINING THE TF MAP() AND TAKE() OPERATORS

Listing 3.21 displays the contents of `tf-map-take.py`, which illustrates how to invoke the `map()` operator followed by the `take()` operator in TensorFlow with a one-shot iterator.

LISTING 3.21: tf-map-take.py

```
import tensorflow as tf
import numpy as np

# a simple Numpy array
x = np.array([[1],[2],[3],[4]])

# make a dataset from a Numpy array
dataset = tf.data.Dataset.from_tensor_slices(x)

tf.enable_eager_execution()

# a simple Numpy array
x = np.array([[1],[2],[3],[4]])

# make a dataset from a numpy array
dataset = tf.data.Dataset.from_tensor_slices(x)
dataset = dataset.map(lambda x: x*2).map(lambda x: x+1).
                                     map(lambda x: x**3)

# define an iterator
iter = dataset.make_one_shot_iterator()
el = iter.get_next()

for value in dataset.take(2):
  print("value:",value)
```

Listing 3.21 initializes the variable `x` as a NumPy array consisting of four elements, where each element is a 1x1 array consisting of the numbers 1, 2, 3, and 4. Next, the variable `dataset` is initialized as a TF `Dataset` that is created from the contents of the variable x.

The next portion of Listing 3.21 involves three lambda expressions that's shown in bold and reproduced here:

```
dataset = dataset.map(lambda x: x*2).map(lambda x: x+1).
                                      map(lambda x: x**3)
```

The preceding code snippet transforms each input value by first doubling the value, then adding one to the first result, and then cubing the second result.

The next code snippet defines the variable `iter` as a "one shot" iterator, followed by the variable `el` that is populated with the first element in the variable `dataset`.

The final portion of Listing 3.21 "takes" only the first two elements from the variable `dataset` and displays their contents, as shown here:

```
('value:', <tf.Tensor: id=36, shape=(1,), dtype=int64,
                                   numpy=array([27])>)
('value:', <tf.Tensor: id=38, shape=(1,), dtype=int64,
                                   numpy=array([125])>)
```

Note that eager execution *must* be enabled; otherwise you will see the following error message:

```
RuntimeError: dataset.__iter__() is only supported when
                            eager execution is enabled.
```

COMBINING THE TF ZIP() AND BATCH() OPERATORS

Listing 3.22 displays the contents of `tf-zip-batch.py` that illustrates how to combine the `zip()` and `batch()` operators in TensorFlow.

LISTING 3.22: tf-zip-batch.py

```
import tensorflow as tf
import numpy as np

ds1 = tf.data.Dataset.range(100)
ds2 = tf.data.Dataset.range(0, -100, -1)
ds3 = tf.data.Dataset.zip((ds1, ds2))
ds4 = ds3.batch(4)

iterator = ds4.make_one_shot_iterator()
next_element = iterator.get_next()

with tf.Session() as sess:
  print(sess.run(next_element))
  print(sess.run(next_element))
  print(sess.run(next_element))
```

Listing 3.22 initializes the variables ds1, ds2, ds3, and ds4 as TF Datasets that are created successively, starting from ds1 that contains the integers between 0 and 99 inclusive.

The next portion of Listing 3.22 defines the variable iterator as a "one shot" iterator, followed by the variable next_element that is populated with the first element in the variable dataset. The final portion of Listing 3.22 prints three lines of "batched" output, as shown here:

```
(array([0, 1, 2, 3]), array([ 0, -1, -2, -3]))
(array([4, 5, 6, 7]), array([-4, -5, -6, -7]))
(array([ 8, 9, 10, 11]), array([ -8, -9, -10, -11]))
```

COMBINING THE TF ZIP() AND TAKE() OPERATORS

Listing 3.23 displays the contents of tf-zip-take.py, which illustrates how to combine the zip() and take() operators in TensorFlow.

LISTING 3.23: tf-zip-take.py

```
import tensorflow as tf
import numpy as np

x = np.arange(0, 10)
y = np.arange(1, 11)

# create dataset objects from the arrays
dx = tf.data.Dataset.from_tensor_slices(x)
dy = tf.data.Dataset.from_tensor_slices(y)

# zip the two datasets together
dcomb = tf.data.Dataset.zip((dx, dy)).batch(3)
iterator = dcomb.make_initializable_iterator()

# extract an element
next_element = iterator.get_next()

with tf.Session() as sess:
  sess.run(iterator.initializer)
  for i in range(15):
    val = sess.run(next_element)
    print(val)
    if (i + 1) % (10 // 3) == 0 and i > 0:
      sess.run(iterator.initializer)
```

Listing 3.23 initializes the variables x and y as a range of integers from 0 to 11 and from 0 to 12, respectively. Next, the variables dx and dy are initialized as TF Datasets that are created from the contents of the variables x and y, respectively.

The next code snippet defines the variable dcomb as a TF Dataset that combines the elements from dx and dy in a pairwise fashion via the zip() operator, as shown here:

```
dcomb = tf.data.Dataset.zip((dx, dy)).batch(3)
```

Notice how method chaining is performed by "tacking on" the batch(3) operator as part of the definition of dcomb.

The next code snippet defines the variable iterator as an "initializable" iterator, followed by the variable next_element that is populated with the first integer in dx (which is the integer 0).

The final portion of Listing 3.23 contains a loop that executes 15 times, and during each iteration the loop prints the current contents of the variable iterator. Each line of output consists of two "blocks" of numbers, where a block consists of three consecutive integers. The output from launching the code in Listing 3.23 is here:

```
(array([0, 1, 2]), array([1, 2, 3]))
(array([3, 4, 5]), array([4, 5, 6]))
(array([6, 7, 8]), array([7, 8, 9]))
(array([0, 1, 2]), array([1, 2, 3]))
(array([3, 4, 5]), array([4, 5, 6]))
(array([6, 7, 8]), array([7, 8, 9]))
(array([0, 1, 2]), array([1, 2, 3]))
(array([3, 4, 5]), array([4, 5, 6]))
(array([6, 7, 8]), array([7, 8, 9]))
(array([0, 1, 2]), array([1, 2, 3]))
(array([3, 4, 5]), array([4, 5, 6]))
(array([6, 7, 8]), array([7, 8, 9]))
(array([0, 1, 2]), array([1, 2, 3]))
(array([3, 4, 5]), array([4, 5, 6]))
(array([6, 7, 8]), array([7, 8, 9]))
```

TF DATASETS AND RANDOM NUMBERS (1)

Listing 3.24 displays the contents of tf-random-one-shot.py, which illustrates how to create a TF Dataset with random numbers.

LISTING 3.24: tf-random-one-shot.py

```
import tensorflow as tf
import numpy as np

x = np.random.sample((100,2))

# make a dataset from a numpy array
dataset = tf.data.Dataset.from_tensor_slices(x)

iter = dataset.make_one_shot_iterator()
el = iter.get_next()

with tf.Session() as sess:
  print(sess.run(el))
```

Listing 3.24 initializes the variable x as a NumPy array consisting of 100 rows and 2 columns of randomly generated numbers. Next, the variable dataset is initialized as a TF Dataset that is created from the contents of the variable x.

The next code snippet defines the variable `iter` as a "one shot" iterator, followed by the variable `el` that is populated with the first element in the variable dataset.

The final portion of Listing 3.24 prints the first line of transformed data, as shown here:

```
[0.69707335 0.21129127]
```

TF DATASETS AND RANDOM NUMBERS (2)

Listing 3.25 displays the contents of `tf-random-one-shot2.py`, which illustrates how to create a TF `Dataset` with random numbers.

LISTING 3.25: tf-random-one-shot2.py

```
import tensorflow as tf
import numpy as np

# two Numpy arrays with random numbers
features, labels = (np.random.sample((100,2)), np.random.
                                      sample((100,1)))
dataset = tf.data.Dataset.from_tensor_
                           slices((features,labels))

iter = dataset.make_one_shot_iterator()
el = iter.get_next()

with tf.Session() as sess:
  print(sess.run(el))
```

Listing 3.25 initializes the variables `features` and `labels` that are generated from a `NumPy` array consisting of 100 rows and 2 columns of randomly generated numbers, and from a `NumPy` array consisting of 100 rows and 1 column of randomly generated numbers.

Next, the variable `dataset` is initialized as a TF `Dataset` that is created from the contents of the variables `features` and `labels`. The next code snippet defines the variable `iter` as a "one shot" iterator, followed by the variable `el` that is populated with the first element in the variable `dataset`. The final portion of Listing 3.25 prints the first line of transformed data, as shown here:

```
(array([0.47582573, 0.36387037]), array([0.26763744]))
```

TF DATASETS FROM CSV FILES

Listing 3.26 displays the contents of `simple.csv`, and Listing 3.27 displays the contents of `tf-csv-dataset.py` that illustrates how to create a TF `Dataset` from data in a CSV file.

LISTING 3.26: *simple.csv*

```
text,sentiment
'a good restaurant', 1
'a poor movie ', 0
'a great dessert', 1
'a plain hamburger', 0
```

LISTING 3.27: *tf-csv-dataset.py*

```
import tensorflow as tf
import numpy as np

csv_file = './simple.csv'
dataset = tf.data.experimental.make_csv_dataset(csv_file,
                                                batch_size=2)
iter = dataset.make_one_shot_iterator()
next = iter.get_next()

# next is a dict with key=columns names and value=column
                                                        data
print("next:",next)
inputs, labels = next['text'], next['sentiment']

with tf.Session() as sess:
  sess.run([inputs, labels])
  print("inputs:",sess.run([inputs, labels]))
```

Listing 3.27 initializes the variable `csv_file` with the name of a CSV file (which is `simple.csv` in this example). Next, the variable `dataset` is initialized as a CSV-based TF `Dataset` that is created from the contents of the variable `csv_file`. The next code snippet defines the variable `iter` as a "one shot" iterator, followed by the variable `next` that is populated with the first element in the variable `dataset`.

The next portion of Listing 3.27 initializes the variables `inputs` and `labels` with the first column and second column, respectively, of the first line of data from the CSV file. The last portion of Listing 3.27 displays the values of `inputs` and `labels`, as shown here:

```
('next:', OrderedDict([('text', <tf.Tensor 'IteratorGetNext:1'
        shape=(2,) dtype=string>), ('sentiment', <tf.Tensor
            'IteratorGetNext:0' shape=(2,)dtype=int32>)]))
('inputs:', [array([""'a poor movie '", "'a good
  restaurant'"], dtype=object), array([0, 1], dtype=int32)])
```

WORKING WITH TF.ESTIMATORS AND TF.LAYERS (OPTIONAL)

The first sub-section introduced below is useful if you have some experience with well-known machine learning algorithms using a Python library such as `scikit-learn`. You will see a list of the TensorFlow classes that are similar to their Python-based counterparts in machine learning, with classes for regression tasks and classes for classification tasks.

The second subsection contains a list of TensorFlow classes that are relevant for defining CNNs in TensorFlow.

If you are new to machine learning, then this section will have much more limited value to you right now, but you can still learn what's available for future reference.

What are TF Estimators?

The `tf.estimator` namespace contains an assortment of classes that implement various algorithms that are available in machine learning, such as boosted trees, DNN classifiers, DNN regressors, linear classifiers, and linear regressors.

The estimator-related classes `DNNRegressor`, `LinearRegressor`, and `DNNLinearCombinedRegressor` are for regression tasks, whereas the classes `DNNClassifier`, `LinearClassifier`, and `DNNLinearCombinedClassifier` are for classification tasks. A more extensive list of estimator classes (with very brief descriptions) is listed below:

- BoostedTreesClassifier: A classifier for Tensorflow Boosted Trees models
- BoostedTreesRegressor: A cegressor for Tensorflow Boosted Trees models
- CheckpointSaverHook: Saves checkpoints every N steps or seconds
- DNNClassifier: A classifier for TensorFlow DNN models
- DNNEstimator: An estimator for TensorFlow DNN models with user-specified head
- DNNLinearCombinedClassifier: An estimator for TensorFlow Linear and DNN joined classification models
- DNNLinearCombinedRegressor: An estimator for TensorFlow Linear and DNN joined models for regression
- DNNRegressor: A regressor for TensorFlow DNN models
- Estimator: Estimator class to train and evaluate TensorFlow models
- LinearClassifier: Linear classifier model
- LinearEstimator: An estimator for TensorFlow linear models with user-specified head
- LinearRegressor: An estimator for TensorFlow Linear regression problems

All estimator classes are in the `tf.estimator` namespace, and all the estimator classes inherit from the `tf.estimator.Estimator` class. Read the online documentation for the details of the preceding classes as well as online tutorials for relevant code samples.

TF Estimators and tf.data.Dataset (optional)

This section is marked "optional" because the code requires TF 2, and it's included here to give you a preview of the functionality in this code sample.

In addition to creating a dataset from NumPy arrays of data or from Pandas Dataframes, you can create a dataset from existing datasets. For example,

Listing 3.28 displays the contents of `tf2-mnist-estimator.py` that illustrates how to create a `tf.data.Dataset` from the MNIST dataset.

LISTING 3.28: tf2-mnist-estimator.py

```
import tensorflow as tf

train, test = tf.keras.datasets.mnist.load_data()
mnist_x, mnist_y = train

mnist_ds = tf.data.Dataset.from_tensor_slices(mnist_x)
print(mnist_ds)

iterator = tf.compat.v1.data.make_one_shot_iterator(mnist_ds)

count = 0
max_count = 4

try:
  while True:
    print("value:",iterator.get_next())
    count += 1
    if(count > max_count):
      break
except tf.errors.OutOfRangeError:
  pass
```

Listing 3.28 initializes the variables train and test from the MNIST dataset, and then initializes the variables `mnist_x` and `mnist_y` from the train variable. The next code snippet initializes the `mnist_ds` variable as a `tf.data.Dataset` that is created from the `mnist_x` variable. Next, the variable `iterator` is defined as a "one shot" iterator that we'll use to iterate through the elements in the dataset.

The next portion of Listing 3.28 contains a `try/except` code block that iterates through the data, and prints the first 5 rows of data. If there are fewer than 5 rows, then the `except` code block is executed, which allows this code to terminate gracefully (i.e., without an error message).

The complete output from launching the code in Listing 3.30 is available in the text file `tf2-mnist-estimator.out`. The next block shows you a portion of the data (i.e., the pixel values) in the first image contained in the MNIST dataset.

```
value: tf.Tensor(
[[ 0    0    0    0    0    0    0    0    0    0    0    0    0    0
                                              0    0    0    0
     0    0    0    0    0    0    0    0    0    0]
 [ 0    0    0    0    0    0    0    0    0    0    0    0    0    0
                                              0    0    0    0
     0    0    0    0    0    0    0    0    0    0]
 [ 0    0    0    0    0    0    0    0    0    0    0    0    0    0
                                              0    0    0    0
     0    0    0    0    0    0    0    0    0    0]
```

```
  [ 0    0    0    0    0    0    0    0    0    0    0    0    0    0
                                                  0    0    0    0
    0    0    0    0    0    0    0    0    0    0]
  [ 0    0    0    0    0    0    0    0    0    0    0    0    0    0
                                                  0    0    0    0
    0    0    0    0    0    0    0    0    0    0]
  [ 0    0    0    0    0    0    0    0    0    0    0    0    3   18
                                                 18   18  126  136
  175   26  166  255  247  127    0    0    0    0]
  [ 0    0    0    0    0    0    0    0   30   36   94  154  170  253
                                                253  253  253  253
  225  172  253  242  195   64    0    0    0    0]
// output omitted for brevity
  [ 0    0    0    0   55  172  226  253  253  253  253  244  133   11
                                                  0    0    0    0
    0    0    0    0    0    0    0    0    0    0]
  [ 0    0    0    0  136  253  253  253  212  135  132   16    0    0
                                                  0    0    0    0
    0    0    0    0    0    0    0    0    0    0]
  [ 0    0    0    0    0    0    0    0    0    0    0    0    0    0
                                                  0    0    0    0
    0    0    0    0    0    0    0    0    0    0]
  [ 0    0    0    0    0    0    0    0    0    0    0    0    0    0
                                                  0    0    0    0
    0    0    0    0    0    0    0    0    0    0]
  [ 0    0    0    0    0    0    0    0    0    0    0    0    0    0
                                                  0    0    0    0
    0    0    0    0    0    0    0    0    0    0]], shape=(28,
                                                 28), dtype=uint8)
```

What are tf.layers?

The tf.layers namespace contains an assortment of classes for the layers in Neural Networks, including DNNs and CNNs . Some of the more common classes in the tf.layers namespace are listed below:

- BatchNormalization: Batch Normalization layer
- Conv2D: 2D convolution layer (e.g., spatial convolution over images)
- Dense: Densely-connected layer class
- Dropout: Applies dropout to the input
- Flatten: Flattens an input tensor while preserving the batch axis (axis 0)
- Layer: Base layer class
- MaxPooling2D: Max pooling layer for 2D inputs (e.g., images)

For example, a minimalistic CNN starts with a "triple" that consists of a Conv2D layer, followed by ReLU (Rectified Linear Unit) activation function, and then a MaxPooling2D layer. If you see this triple appear a second time, followed by two consecutive Dense layers and then a softmax activation function, it's known as "LeNet."

A bit of trivia: In the late 1990s, when people deposited cheques at an automated bank machine, LeNet scanned the contents of those cheques to determine the digits of the cheque amount (of course, customers had to confirm that the number determined by LeNet was correct). LeNet had an accuracy rate around 90%, which is a very impressive result for such a simple convolutional neural network!

Other Useful Parts of TensorFlow

In addition to the classes in the previous sections, TensorFlow provides a number of other useful namespaces, including the following list:

- tf.data
- tf.linalg
- tf.lite
- tf.losses
- tf.math
- tf.nn
- tf.random
- tf.saved_model
- tf.test
- tf.train
- tf.version

The `tf.data` namespace contains the `tf.data.Dataset` namespace, which contains classes that are discussed in the first half of this chapter; the `tf.linalg` namespace contains an assortment of classes that perform operations in linear algebra; the `tf.lite` namespace contains classes for mobile application development.

The `tf.math` namespace contains classes for trigonometric calculations; the `tf.nn` namespace contains classes for batch normalization, RNNs, dropout rate, and max pooling. Read the extensive online documentation for more details regarding these (and other classes) in TensorFlow.

WHAT IS A TFRECORD?

A `TFRecord` is a file that describes the data required during the training phase and the testing phase of a model. There are two protocol buffer message types available for a `TFRecord`: the `Example` message type and the `SequenceExample` message type. These protocol buffer message types enable you to arrange data as a map from string keys to values that are lists of integers, 32-bit floats, or bytes.

The data in a `TFRecord` is "wrapped" inside a `Feature` class. In addition, each feature is stored in a key value pair, where the key corresponds to the title that is allotted to each feature. These titles are used later for extracting the data from `TFRecord`. The created dictionary is passed as input to a `Feature` class. Finally, the features object is passed as input to `Example` class that is appended into the TFRecord. The preceding process is repeated for every type of data that is stored in TFRecord.

The TFRecord file format is a record-oriented binary format that you can use for training data. In addition, the tf.data.TFRecordDataset class enables you to stream over the contents of one or more TFRecord files as part of an input pipeline.

You can store any type of data, including images, in the Example format. However, you specify the mechanism for arranging the data into serialized bytes, as well as reconstructing the original format.

A Simple TFRecord

Listing 3.29 displays the contents of tf-record1.py that illustrates how to define a TFRecord.

LISTING 3.29: tf-record1.py

```
import tensorflow as tf

simple1 = tf.train.Example(features=tf.train.
                                     Features(feature={
  'my_ints': tf.train.Feature(int64_list=tf.train.
                             Int64List(value=[2, 5])),
  'my_float': tf.train.Feature(float_list=tf.train.
                             FloatList(value=[3.6])),
  'my_bytes': tf.train.Feature(bytes_list=tf.train.
                             BytesList(value=['data']))
}))

print("my_ints:",  simple1.features.feature['my_ints'].
                                      int64_list.value)
print("my_floats:",simple1.features.feature['my_float'].float_
                                           list.value)
print("my_bytes:",  simple1.features.feature['my_bytes'].
                                      bytes_list.value)

#print("simple1:",simple1)
```

Listing 3.29 contains the definition of the variable simple1 that is an instance of the tf.train.Example class. The simple1 variable defines a record consisting of the fields my_ints, my_floats, and my_bytes that are of type Int64List, FloatList, and ByteList, respectively. The final portion of Listing 3.29 contains print() statements that display the values of various elements in the simple1 variable, as shown here:

```
('my_ints:', [2L, 5L])
('my_floats:', [3.5999999046325684])
('my_bytes:', ['data'])
```

SUMMARY

This chapter introduced you to TensorFlow `Datasets` and iterators that are well-suited for processing the contents of "normal" size datasets as well datasets that are too large to fit in memory. You saw how to define a lambda expression and use that expression in a TensorFlow `Dataset`.

Next, you learned about various "lazy operators," including `filter()`, `map()`, `filter()`, `flatmap()`, `take()`, and `zip()`, and how to use them to define a subset of the data in a TensorFlow `Dataset`. You also learned how to use one-shot iterators and reusable iterators in TensorFlow in order to iterate through the elements of a TensorFlow `Datasets`.

Then you got a brief introduction to the `tf.estimators` namespace, which contains an assortment of classes that implement various algorithms, such as boosted trees, DNN classifiers, DNN regressors, linear classifiers, and linear regressors. After that, you learned that the `tf.layers` namespace contains an assortment of classes for DNNs and CNNs.

Finally, you learned about a TensorFlow `TFRecord`, which is a file that describes the data required during the training phase and the testing phase of a model.

## CHAPTER 4

LINEAR REGRESSION

This chapter introduces linear regression, along with code samples of linear regression: first with NumPy APIs, and then with TensorFlow APIs. The code samples use an "incremental" approach, starting with simple examples that involve Python and NumPy code (often using the NumPy linspace() API). Then you will see comparable code samples involving Tensor-Flow code. In addition, the samples in the second half of the chapter usually involve concepts from the first half of the chapter.

The first part of this chapter briefly discusses the basic concepts involved in linear regression. Although linear regression was developed more than 200 years ago, this technique is still one of the "core" techniques for solving (albeit simple) problems in statistics and machine learning. In fact, the technique known as "mean squared error" (MSE) for finding a best-fitting line for data points in a 2D plane (or a hyperplane for higher dimensions) is implemented in Python and TensorFlow in order to minimize so-called "cost" functions that are discussed later.

The second section in this chapter contains additional code samples involving linear regression tasks using standard techniques in NumPy. Hence, if you are comfortable with this topic, you can probably skim quickly through the first two sections of this chapter. The third section shows you how to solve linear regression using TensorFlow.

WHAT IS LINEAR REGRESSION?

The goal of linear regression is to find the best-fitting line that "represents" a dataset. Keep in mind two key points. First, the best-fitting line does not necessarily pass through all (or even most of) the points in the dataset. The purpose of a best-fitting line is to minimize the distance of that line from the points in the dataset. Second, linear regression does not determine the best-fitting polynomial: the latter involves finding a higher-degree polynomial that passes through many of the points in a dataset.

Moreover, a dataset in the plane can contain two or more points that lie on the same *vertical* line, which is to say that those points have the same x value. However, a function *cannot* pass through such a pair of points: if two points (x1,y1) and (x2,y2) have the same x value, then they must have the same y value (i.e., y1=y2). On the other hand, a function can have two or more points that lie on the same *horizontal* line.

Now consider a scatter plot with many points in the plane that are sort of "clustered" in an elongated cloud-like shape: a best-fitting line will probably intersect only limited number of points (in fact, a best-fitting line might not intersect *any* of the points).

One other scenario to keep in mind: suppose a dataset contains a set of points that lie on the same line. For instance, let's say the x values are in the set {1,2,3,...,10} and the y values are in the set {2,4,6,...,20}. Then the equation of the best-fitting line is y=2*x+0. In this scenario, all the points are *collinear*, which is to say that they lie on the same line.

Linear Regression versus Curve-Fitting

Suppose a dataset consists of n data points of the form (x, y), and no two of those data points have the same x value. Then there is a polynomial of degree less than or equal to n-1 that passes through those n points (if you are really interested, you can find a mathematical proof of this statement in online articles). For example, a line is a polynomial of degree one and it can intersect any pair of non-vertical points in the plane. For any triple of points (that are not all on the same line) in the plane, there is a quadratic equation that passes through those points.

In addition, sometimes a lower degree polynomial is available. For instance, consider the set of 100 points in which the x value equals the y value: in this case, the line y = x (which is a polynomial of degree one) passes through all 100 points.

However, keep in mind that the extent to which a line "represents" a set of points in the plane depends on how closely those points can be approximated by a line, which is measured by the *variance* of the points (the variance is a statistical quantity). The more collinear the points, the smaller the variance; conversely, the more "spread out" the points are, the larger the variance.

When are Solutions Exact Values?

Although statistics-based solutions provide closed-form solutions for linear regression, neural networks provide *approximate* solutions. This is due to the fact that machine learning algorithms for linear regression involve a sequence of approximations that "converges" to optimal values, which means that machine learning algorithms produce estimates of the exact values. For example, the slope m and y-intercept b of a best-fitting line for a set of points a 2D plane have a closed-form solution in statistics, but they can only be approximated via machine learning algorithms (exceptions do exist, but they are rare situations).

Keep in mind that even though a closed-form solution for "traditional" linear regression provides an exact value for both m and b, sometimes you can only use an approximation of the exact value. For instance, suppose that the slope m of a best-fitting line equals the square root of 3 and the y-intercept b is the square root of 2. If you plan to use these values in source code, you can only work with an approximation of these two numbers. In the same scenario, a neural network computes approximations for m and b, regardless of whether or not the exact values for m and b are irrational, rational, or integer values. However, machine learning algorithms are better suited for complex, non-linear, multidimensional datasets, which is beyond the capacity of linear regression.

As a simple example, suppose that the closed form solution for a linear regression problem produces integer or rational values for both m and b. Specifically, let's suppose that a closed form solution yields the values 2.0 and 1.0 for the slope and y-intercept, respectively, of a best-fitting line. The equation of the line looks like this:

```
y = 2.0 * x + 1.0
```

However, the corresponding solution from training a neural network might produce the values 2.0001 and 0.9997 for the slope m and the y-intercept b, respectively, as the values of m and b for a best-fitting line. Always keep this point in mind, especially when you are training a neural network.

What is Multivariate Analysis?

Multivariate analysis generalizes the equation of a line in the Euclidean plane to higher dimensions, and it's called a *hyper plane* instead of a line. The generalized equation has the following form:

```
y = w1*x1 + w2*x2 + . . . + wn*xn + b
```

In the case of 2D linear regression, you only need to find the value of the slope (m) and the y-intercept (b), whereas in multivariate analysis you need to find the values for w1, w2, . . ., wn. Note that multivariate analysis is a term from statistics, and in machine learning it's often referred to as "generalized linear regression."

Keep in mind that most of the code samples in this book that pertain to linear regression involve 2D points in the Euclidean plane.

OTHER TYPES OF REGRESSION

Linear regression finds the best fitting line that "represents" a dataset, but what happens if a line in the plane is not a good fit for the dataset? This is a relevant question when you work with datasets.

Some alternatives to linear regression include quadratic equations, cubic equations, or higher-degree polynomials. However, these alternatives involve trade-offs, as we'll discuss later.

Another possibility is a sort of hybrid approach that involves piece-wise linear functions, which comprise a set of line segments. If contiguous line segments are connected, then it's a piece-wise linear continuous function; otherwise it's a piece-wise linear discontinuous function.

Thus, given a set of points in the plane, regression involves addressing the following questions:

1. What type of curve fits the data well? How do we know?
2. Does another type of curve fit the data better?
3. What does "best fit" mean?

One way to check if a line fits the data involves a visual check, but this approach does not work for data points that are higher than two dimensions. Moreover, this is a subjective decision, and some sample datasets are displayed later in this chapter. By visual inspection of a dataset, you might decide that a quadratic or cubic (or even higher degree) polynomial has the potential of being a better fit for the data. However, visual inspection is probably limited to points in a 2D plane or in three dimensions.

Let's defer the non-linear scenario and let's make the assumption that a line would be a good fit for the data. There is a well-known technique for finding the "best fitting" line for such a dataset, and it's called MSE that we'll discuss later in this chapter.

The next section provides a quick review of linear equations in the plane, along with some images that illustrate examples of linear equations.

WORKING WITH LINES IN THE PLANE (OPTIONAL)

This section contains a short review of lines in the Euclidean plane, so you can skip this section if you are comfortable with this topic. A minor point that's often overlooked is that lines in the Euclidean plane have infinite length. If you select two distinct points of a line, then all the points between those two selected points is a *line segment*. A *ray* is a "half infinite" line: when you select one point as an endpoint, then all the points on one side of the line constitutes a ray.

For example, the points in the plane for which y-coordinate is 0 is a line (it's the x-axis), whereas the points between (0,0) and (1,0) on the x-axis form a line segment. In addition, the points on the x-axis that are to the right of (0,0) form a ray, and the points on the x-axis that are to the left of (0,0) also form a ray.

For simplicity and convenience, in this book we'll use the terms "line" and "line segment" interchangeably, and now let's delve into the details of lines in the Euclidean plane. Just in case you're a bit fuzzy on the details, here is the equation of a (non-vertical) line in the Euclidean plane:

```
y = m*x + b
```

The value of m is the slope of the line and the value of b is the y-intercept (i.e., the place where the line intersects the y-axis).

If need be, you can use a more general equation that can also represent vertical lines, as shown here:

```
a*x + b*y + c = 0
```

However, we won't be working with vertical lines, so we'll stick with the first formula.

Figure 4.1 displays three horizontal lines whose equations (from top to bottom) are $y = 3$, $y = 0$, and $y = -3$, respectively.

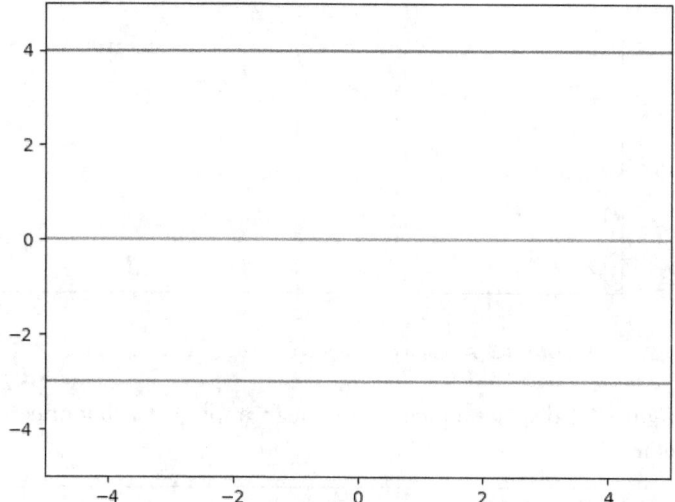

FIGURE 4.1 A Graph of Three Horizontal Line Segments.

Figure 4.2 displays two slanted lines whose equations are $y = x$ and $y = -x$, respectively.

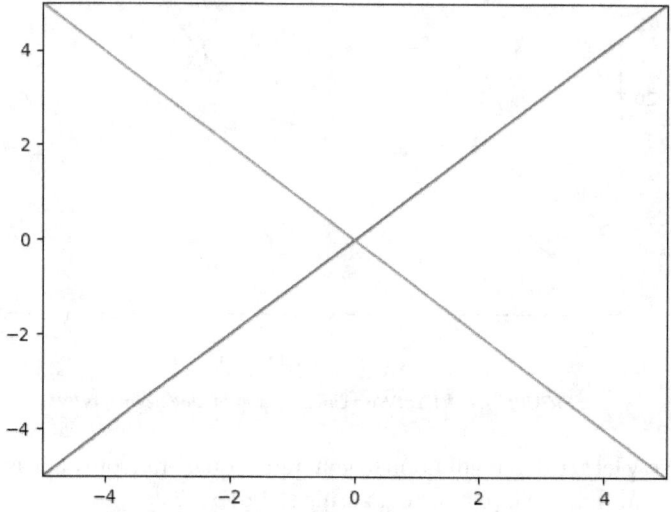

FIGURE 4.2 A Graph of Two Diagonal Line Segments.

Figure 4.3 displays two slanted parallel lines whose equations are y = 2*x and y = 2*x + 3, respectively.

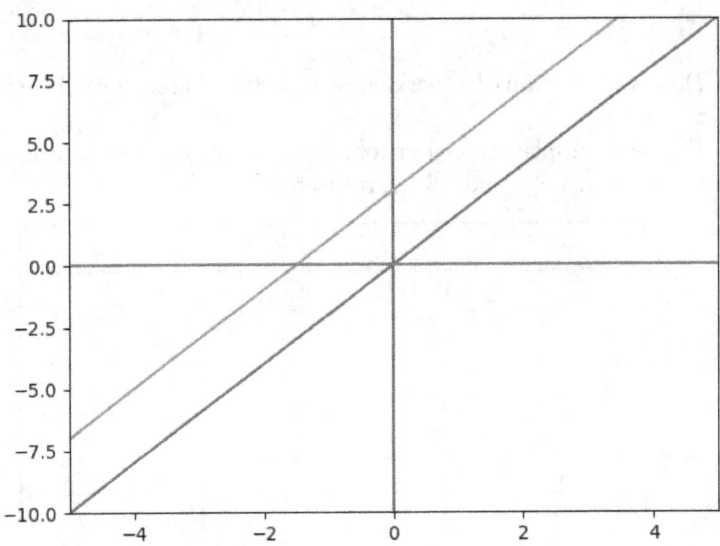

FIGURE 4.3 A Graph of Two Slanted Parallel Line Segments.

Figure 4.4 displays a piece-wise linear graph consisting of connected line segments.

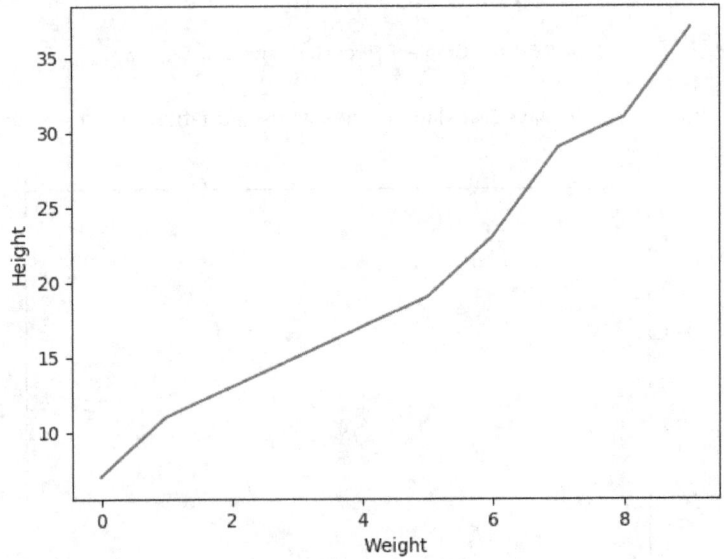

FIGURE 4.4 A PieceWise Linear Graph of Line Segments.

Now let's turn our attention to generating quasi-random data using a NumPy API, and then we'll plot the data using Matplotlib.

SCATTER PLOTS WITH NUMPY AND MATPLOTLIB

Listing 4.1 displays the contents of np-plot1.py, which illustrates how to use the NumPy randn() API to generate a dataset and then the scatter() API in Matplotlib to plot the points in the dataset.

One detail to note is that all the adjacent horizontal values are equally spaced, whereas the vertical values are based on a linear equation plus a "perturbation" value. This "perturbation technique" (which is not a standard term) is used in other code samples in this chapter in order to add a slightly randomized effect when the points are plotted. The advantage of this technique is that the best-fitting values for m and b are known in advance, and therefore we do not need to guess their values.

LISTING 4.1: np-plot1.py

```
import numpy as np
import matplotlib.pyplot as plt

x = np.random.randn(15,1)
y = 2.5*x + 5 + 0.2*np.random.randn(15,1)

print("x:",x)
print("y:",y)

plt.scatter(x,y)
plt.show()
```

Listing 4.1 contains two import statements and then initializes the array variable x with 15 random numbers between 0 and 1.

Next, the array variable y is defined in two parts: the first part is a linear equation 2.5*x + 5 and the second part is a "perturbation" value that is based on a random number. Thus, the array variable y simulates a set of values that closely approximate a line segment.

This technique is used in code samples that simulate a line segment, and then the training portion approximates the values of m and b for the best-fitting line. Obviously, we already *know* the equation of the best fitting-line: the purpose of this technique is to compare the trained values for the slope m and y-intercept b with the known values (which in this case are 2.5 and 5).

A partial output from Listing 4.1 is here:

```
x: [[-1.42736308]
 [ 0.09482338]
 [-0.45071331]
 [ 0.19536304]
 [-0.22295205]
 // values omitted for brevity
y: [[1.12530514]
 [5.05168677]
 [3.93320782]
 [5.49760999]
```

```
[4.46994978]
// values omitted for brevity
```

Figure 4.5 displays a scatter plot of points based on the values of x and y.

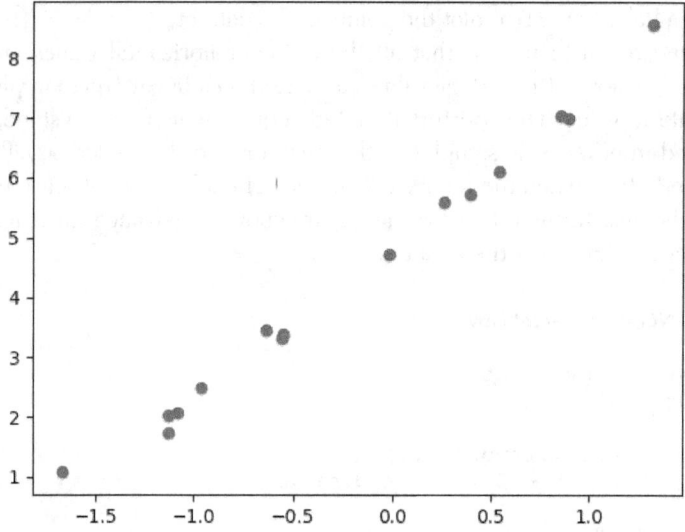

FIGURE 4.5 A Scatter Plot of Points for a Line Segment.

Why the "Perturbation Technique" is Useful

You already saw how to use the "perturbation technique" and, by way of comparison, consider a dataset with the following points that are defined in the Python array variables X and Y:

```
X = [0,0.12,0.25,0.27,0.38,0.42,0.44,0.55,0.92,1.0]
Y = [0,0.15,0.54,0.51, 0.34,0.1,0.19,0.53,1.0,0.58]
```

If you need to find the best fitting line for the preceding dataset, how would you guess the values for the slope m and the y-intercept b? In most cases, you probably cannot guess their values. On the other hand, the "perturbation technique" enables you to "jiggle" the points on a line whose value for the slope m (and optionally the value for the y-intercept b) is specified in advance.

Keep in mind that the "perturbation technique" only works when you introduce small random values that do not result in different values for m and b.

SCATTER PLOTS WITH NUMPY AND MATPLOTLIB

The code in Listing 4.1 assigned random values to the variable x, whereas a hard-coded value is assigned to the slope m. The y values are a hard-coded multiple of the x values, plus a random value that is calculated via the "perturbation technique." Hence we do not know the value of the y-intercept b.

In this section, the values for `trainX` are based on the `np.linspace()` API, and the values for `trainY` involve the "perturbation technique" that is described in the previous section.

The code in this example simply prints the values for `trainX` and `trainY`, which correspond to data points in the Euclidean plane. Listing 4.2 displays the contents of `np-plot2.py` that illustrates how to simulate a linear dataset in NumPy.

LISTING 4.2: np-plot2.py

```
import numpy as np

trainX = np.linspace(-1, 1, 11)
trainY = 4*trainX + np.random.randn(*trainX.shape)*0.5

print("trainX: ",trainX)
print("trainY: ",trainY)
```

Listing 4.2 initializes the NumPy array variable `trainX` via the NumPy linspace() API, followed by the array variable `trainY` that is defined in two parts. The first part is the linear term `4*trainX` and the second part involves the "perturbation technique" that is a randomly generated number. The output from Listing 3.6 is here:

```
trainX:  [-1.  -0.8 -0.6 -0.4 -0.2  0.   0.2  0.4  0.6  0.8
                                                      1. ]
trainY:  [-3.60147459 -2.66593108 -2.26491189 -1.65121314
  -0.56454605  0.22746004 0.86830728  1.60673482  2.51151543
                               3.59573877  3.05506056]
```

The next section contains an example that is similar to Listing 3.2, using the same "perturbation technique" to generate a set of points that approximate a quadratic equation instead of a line segment.

A QUADRATIC SCATTERPLOT WITH NUMPY AND MATPLOTLIB

Listing 4.3 displays the contents of `np-plot-quadratic.py`, which illustrates how to plot a quadratic function in the plane.

LISTING 4.3: np-plot-quadratic.py

```
import numpy as np
import matplotlib.pyplot as plt

#see what happens with this set of values:
#x = np.linspace(-5,5,num=100)

x = np.linspace(-5,5,num=100)[:,None]
y = -0.5 + 2.2*x +0.3*x**2 + 2*np.random.randn(100,1)
print("x:",x)
```

```
plt.plot(x,y)
plt.show()
```

Listing 4.3 initializes the array variable x with the values that are generated via the np.linspace() API, which in this case is a set of 100 equally spaced decimal numbers between -5 and 5. Notice the snippet [:,None] in the initialization of x, which results in an array of elements, each of which is an array consisting of a single number.

The array variable y is defined in two parts: the first part is a quadratic equation -0.5 + 2.2*x +0.3*x**2 and the second part is a "perturbation" value that is based on a random number (similar to the code in Listing 4.1). Thus, the array variable y simulates a set of values that approximates a quadratic equation. The output from Listing 4.3 is here:

```
x:
[[-5.        ]
 [-4.8989899 ]
 [-4.7979798 ]
 [-4.6969697 ]
 [-4.5959596 ]
 [-4.49494949]
 // values omitted for brevity
 [ 4.8989899 ]
 [ 5.        ]]
```

Figure 4.6 displays a scatter plot of points based on the values of x and y, which have an approximate shape of a quadratic equation.

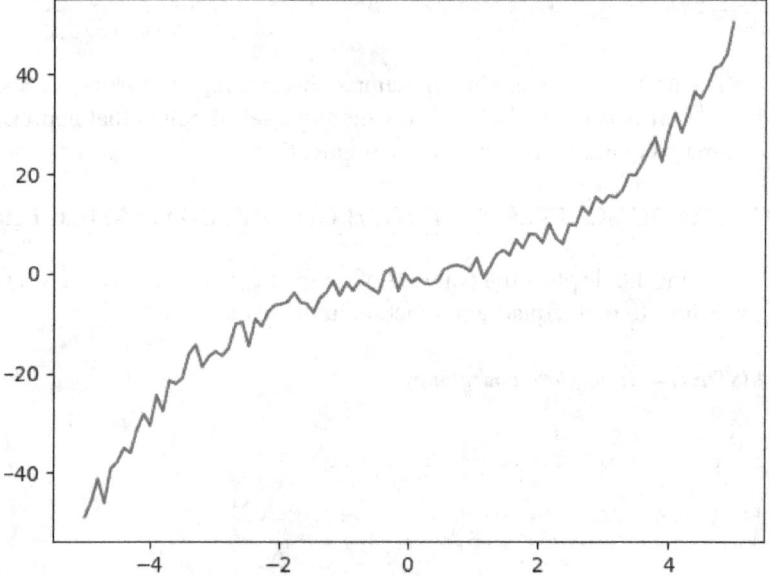

FIGURE 4.6 A Scatter Plot of Points for a Quadratic Equation.

THE MSE FORMULA

Figure 3.8 displays the formula for the MSE. In plain English, the MSE is the sum of the squares of the difference between an actual y value and the predicted y value, divided by the number of points. Notice that the predicted y value is the y value that each point would have if that point were actually on the best-fitting line.

Although the MSE is popular for linear regression, there are other error types available, some of which are discussed briefly in the next section.

A List of Error Types

Although we will only discuss MSE for linear regression in this book, there are other types of formulas that you can use for linear regression, some of which are listed here:

- MSE
- RMSE
- RMSPROP
- MAE

The MSE is the basis for the preceding error types. For example, RMSE is "Root Mean Squared Error," which is the square root of MSE.

On the other hand, MAE is "Mean Absolute Error," which is the sum of *the absolute value of the differences of the y terms* (*not* the square of the differences of the y terms), which is then divided by the number of terms.

The RMSProp optimizer utilizes the magnitude of recent gradients to normalize the gradients. Maintain a moving average over the (root mean squared (RMS) gradients, and then divide that term by the current gradient.

Although it's easier to compute the derivative of MSE, it's also true that MSE is more susceptible to outliers, whereas MAE is less susceptible to outliers. The reason is simple: a squared term can be significantly larger than the absolute value of a term. For example, if a difference term is 10, then a squared term of 100 is added to MSE, whereas only 10 is added to MAE. Similarly, if a difference term is -20, then a squared term 400 is added to MSE, whereas only 20 (which is the absolute value of -20) is added to MAE.

Non-Linear Least Squares

When predicting housing prices, where the dataset contains a wide range of values, techniques such as linear regression or random forests can cause the model to overfit the samples with the highest values in order to reduce quantities such as mean absolute error.

In this scenario, you probably want an error metric, such as relative error, which reduces the importance of fitting the samples with the largest values. This technique is called *non-linear least squares*, which may use a log-based transformation of labels and predicted values.

The next section contains several code samples, the first of which involves calculating the MSE manually, followed by an example that uses `NumPy` formulas to perform the calculations. Finally, we'll look at a TensorFlow example for calculating the MSE.

CALCULATING THE MSE MANUALLY

This section contains two line graphs, both of which contain a line that approximates a set of points in a scatter plot. The MSE for the line in Figure 4.7 is computed as follows:

```
MSE = 1*1 + (-1)*(-1) + (-1)*(-1) + 1*1 = 4
```

Figure 4.7 displays a line segment that approximates a scatter plot of points (some of which intersect the line segment).

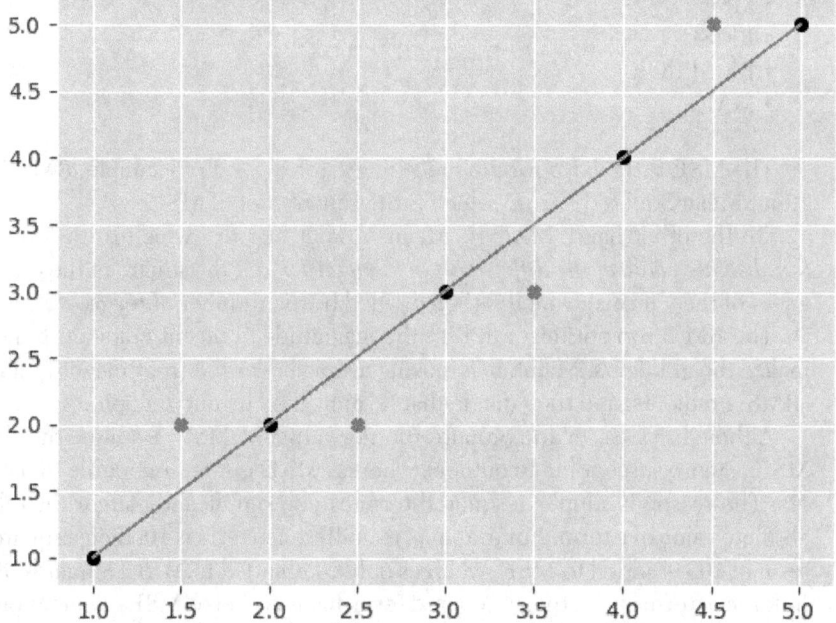

FIGURE 4.7 A Line Graph that Approximates Points of a Scatter Plot.

The MSE for the line in Figure 4.8 is computed as follows:

```
MSE = (-2)*(-2) + 2*2 = 8
```

Figure 4.8 displays a set of points and a line that is a potential candidate for best-fitting line for the data.

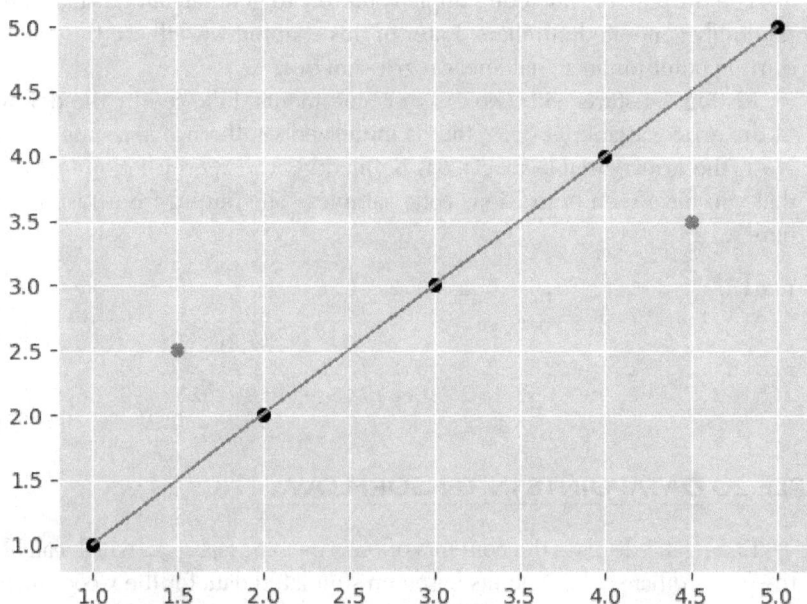

FIGURE 4.8 A Line Graph that Approximates Points of a Scatter Plot.

Thus, the line in Figure 4.7 has a smaller MSE than the line in Figure 4.8, which might have surprised you (or did you guess correctly?)

In these two figures, we calculated the MSE easily and quickly, but in general, it's significantly more difficult. For instance, if we plot 10 points in the Euclidean plane that do not closely fit a line, with individual terms that involve non-integer values, we would probably need a calculator.

A better solution involves NumPy functions, such as the np.linspace() API, as discussed in the next section.

APPROXIMATING LINEAR DATA WITH NP.LINSPACE()

Listing 4.4 displays the contents of tf-linspace1.py, which illustrates how to generate some data with the np.linspace() API in conjunction with the "perturbation technique."

LISTING 4.4: tf-linspace1.py

```
import tensorflow as tf # tf-linspace1.py
import numpy as np

trainX = np.linspace(-1, 1, 6)
trainY = 3*trainX+ np.random.randn(*trainX.shape)*0.5

with tf.Session() as sess:
 print("trainX: ", trainX)
 print("trainY: ", trainY)
```

The purpose of this code sample is merely to generate and display a set of randomly generated numbers. Later in this chapter, we will use this code as a starting point for an actual linear regression task.

Listing 4.4 starts with two `import` statements, followed by the definition of the array variable `trainX` that is initialized via the `np.linspace()` API. Next, the array variable `trainY` is defined via the "perturbation technique" that you have seen in previous code samples. The output from Listing 4.4 is here:

```
trainX:   [-1.   -0.6 -0.2   0.2   0.6   1. ]
trainY:   [-2.9008553   -2.26684745 -0.59516253   0.66452207
                                     1.82669051   2.30549295]
trainX:   [-1.   -0.6 -0.2   0.2   0.6   1. ]
trainY:   [-2.9008553   -2.26684745 -0.59516253   0.66452207
                                     1.82669051   2.30549295]
```

SIMPLE 2D DATAPOINTS IN TENSORFLOW

Listing 4.5 displays the contents of `basic-linreg1.py`, which calculates the y-coordinates of 2D points based on simulated data for the x-coordinates.

LISTING 4.5: basic-linreg1.py

```
import tensorflow as tf

W = tf.Variable([.5], dtype=tf.float32)
b = tf.Variable([-1], dtype=tf.float32)
x = tf.placeholder(tf.float32)
linear_model = W * x + b

with tf.Session() as sess:
  init = tf.global_variables_initializer()
  sess.run(init)
  print(sess.run(linear_model, {x: [1, 2, 3, 4]}))
```

Listing 4.5 contains the variables `W` and `b` and the placeholder `x`, followed by the variable `linear_model` that is the equation of a line in the Euclidean plane. The `with` code block starts with the initialization of the variables (shown in bold). This pair of statements is required whenever TensorFlow variables are defined (such as `W` and `b`) in the code. Finally, a `print()` statement generates the output, as shown here:

```
[-0.5   0.    0.5   1. ]
```

The preceding four values are computed by substituting the x values 1, 2, 3, and 4 in the variable `linear_model`, which, in turn, calculates the four numbers in the preceding array. Now that we know how to generate (x, y) values for a linear equation, let's learn how to calculate the MSE, which is discussed in the next section.

SIMPLE LINEAR MODELS IN TENSORFLOW

The code sample defines a linear equation in TensorFlow based on a hard-coded value 0.5 for the slope (W) and -1 for the y-intercept (b), and then calculates the MSE for a set of values for the variables x and y. Note that the TensorFlow-based linear regression appears later in this chapter.

Listing 4.6 displays the contents of basic-linreg2.py, which illustrates how to compute the MSE with simulated data.

LISTING 4.6: basic-linreg2.py

```
import tensorflow as tf

W = tf.Variable([.5], dtype=tf.float32)
b = tf.Variable([-1], dtype=tf.float32)
x = tf.placeholder(tf.float32)
linear_model = W * x + b

y = tf.placeholder(tf.float32)
squared_deltas = tf.square(linear_model - y)
loss = tf.reduce_sum(squared_deltas)

with tf.Session() as sess:
  init = tf.global_variables_initializer()
  sess.run(init)
  print(sess.run(loss, {x: [1,2,3,4], y: [0,-1,-2,-3]}))
```

Listing 4.6 contains the variables W and b and the placeholder x, followed by the variable linear_model that is the equation of a line in the Euclidean plane. The next block of code is the standard code for initializing a session and the values of variables, followed by a print() statement. The output from Listing 4.6 is here:

```
23.5
```

The next example generates a set of data values using the np.linspace() method and the np.random.randn() method in order to introduces some randomness in the data points.

GRADIENT DESCENT WITH TENSORFLOW (1)

Now we are ready to perform gradient descent with TensorFlow code to find the best-fitting line for a set of data points (at last!). Listing 4.7 displays the contents of tf-linear-regression2.py, which illustrates how to simulate a linear dataset in NumPy in conjunction with the "perturbation technique."

LISTING 4.7: tf-linear-regression2.py

```
import tensorflow as tf
import numpy as np
```

```
trainX = np.linspace(-1, 1, 11)
trainY = 4*trainX + np.random.randn(*trainX.shape)*0.5
print("trainX: ",trainX)
print("trainY: ",trainY)

# Tensorflow stuff:
X = tf.placeholder("float")
Y = tf.placeholder("float")

w = tf.Variable(0.0, name="weights")
y_model = tf.multiply(X, w)

cost = tf.pow(Y-y_model, 2)
train_op = tf.train.GradientDescentOptimizer(0.01).
                                         minimize(cost)

with tf.Session() as sess:
  sess.run(tf.global_variables_initializer())

  for i in range(10):
    for (x, y) in zip(trainX, trainY):
      sess.run(train_op, feed_dict={X: x, Y: y})
      print("w: ",sess.run(w))
```

Listing 4.7 extends the code in Listing 3.6 by defining X, Y, and w, followed by the TensorFlow variable y_model that is the product of w and x. The next portion of Listing 4.7 defines a cost variable that is the MSE of the actual values Y and the values y_model (which represents the best fitting line).

Next, the variable train_op is defined as a gradient descent optimizer, with a learning rate of 0.01, that invokes the minimize() method to minimize the cost variable (defined in the preceding line). This two-line block of code is very common in linear regression; even though the learning_rate might be different, and the gradient descent optimizer might be replaced with an Adam optimizer, the basic structure of these two lines of code is the same. The output from Listing 4.7 is here:

```
trainX:  [-1.   -0.8 -0.6 -0.4 -0.2  0.   0.2  0.4  0.6  0.8
                                                         1. ]
trainY:  [-4.63395653 -3.45196491 -2.13340536 -1.14132441
             -1.13895526  0.11925809 0.88634659  1.77615663
                          2.26356521  3.09042627  3.89392414]
w:   0.34593296
w:   0.6625242
w:   0.9522624
w:   1.2174252
w:   1.4600972
w:   1.6821858
w:   1.8854373
w:   2.0714488
w:   2.2416832
w:   2.3974779
```

Notice two things about the output. First, only the value of w (which is the slope) is displayed, because we are trying to fit a line that passes through the origin to the data points; hence, the value of the y-intercept is 0. Second, the definition of the variable `trainY` is four times the values of `trainX` plus a randomized value for a "perturbation" effect.

We already know that the slope of the best-fitting line is 4, but the computed value is only 2.3974779 because the number of iterations is small. We can improve the accuracy of w by increasing the number of iterations from 10 to 100, and we get the following set of numbers:

```
[lines omitted for brevity]
w:   3.662091
w:   3.6621537
w:   3.6622107
w:   3.6622627
w:   3.662311
```

If we increase the number of iterations from 100 to 1,000, we get the following set of numbers:

```
[lines omitted for brevity]
w:   4.085785
w:   4.085785
w:   4.085785
w:   4.085785
w:   4.085785
```

This time the estimated values for the slope are much closer to the value 4, but notice that the last four calculated values are the same: in fact, *the last 800 calculated values are all equal*! This example shows you that increasing the number of iterations does not always increase the accuracy of the generated approximations.

GRADIENT DESCENT WITH TENSORFLOW (2)

Listing 4.8 displays the contents of `tf-linear-regression3.py` that illustrates how to simulate a linear dataset in NumPy.

LISTING 4.8: tf-linear-regression3.py

```python
import tensorflow as tf
import numpy as np

# model parameters:
W = tf.Variable([.3],  dtype=tf.float32)
b = tf.Variable([-.3], dtype=tf.float32)

# model input and output:
x = tf.placeholder(tf.float32)
linear_model = W * x + b
```

```
y = tf.placeholder(tf.float32)

# loss:
loss = tf.reduce_sum(tf.square(linear_model - y))
optimizer = tf.train.GradientDescentOptimizer(0.01)
train = optimizer.minimize(loss)

x_train = [1,2,3,4]
y_train = [0,-1,-2,-3] # y = -x + 1

init = tf.global_variables_initializer()

with tf.Session() as sess:
  sess.run(init)

  for i in range(100):
    sess.run(train, {x: x_train, y: y_train})
  currW, currB, currLoss =
          sess.run([W, b, loss], {x: x_train, y: y_
                                                  train})

  if( i % 10 == 0):
    print("W: %s b: %s loss: %s" % (currW, currB,
                                              currLoss))
```

Listing 4.8 starts with familiar code: first it initializes the TF variables W and b, followed by the TF placeholder x, and the variable linear_model that is a linear combination of W and b.

Next, the TF placeholder y is defined, followed by the TF variables loss, optimizer, and train that are required in order to train the linear model. If you look carefully, you can see that the loss variable is defined as the MSE. The optimizer variable is gradient descent, and the train variable is the result of minimizing the loss variable.

The next portion of Listing 4.8 initializes the array variables x_train and y_train, whose values represent points on the line y = -x + 1 (which has slope -1 and y-intercept 1). The with code block initializes the TF variables, followed by a loop that executes 100 times. Every iteration of the loop calculates new values for the slope, the y-intercept, and the loss function. The final code snippet prints the values of these three variables after every tenth iteration through the loop. The output from Listing 4.8 is here:

```
W: [-0.21999997] b: [-0.45600003] loss: 4.018144
W: [-0.53481066] b: [-0.36769977] loss: 1.249657
W: [-0.5876153]  b: [-0.2124613]  loss: 0.9820742
W: [-0.6344227]  b: [-0.07484178] loss: 0.7717872
W: [-0.6759173]  b: [0.04715735]  loss: 0.6065279
W: [-0.7127021]  b: [0.15530905]  loss: 0.47665495
W: [-0.7453116]  b: [0.2511851]   loss: 0.37459117
W: [-0.7742198]  b: [0.33617878]  loss: 0.29438165
W: [-0.7998468]  b: [0.41152528]  loss: 0.23134711
W: [-0.8225651]  b: [0.47831967]  loss: 0.18180983
```

If we increase the number of iterations from 100 to 1,000 we get this output:

```
[lines omitted for brevity[
W: [-0.9999944]  b: [0.99998343] loss: 1.8181368e-10
W: [-0.999995]   b: [0.9999853]  loss: 1.4469848e-10
W: [-0.9999956]  b: [0.99998707] loss: 1.1204193e-10
W: [-0.9999961]  b: [0.99998856] loss: 8.744294e-11
W: [-0.99999654] b: [0.99998975] loss: 7.015899e-11
```

We know that the best-fitting line has a slope equal to -1 and y-intercept of 1, and the corresponding approximations are -0.99999654 and 0.99998975, respectively. Thus, the estimated values for the slope and the y-intercept are very close to the correct result, and the value for the loss is essentially equal to zero.

GRADIENT DESCENT WITH FOUR DATAPOINTS

The code sample in this section uses four hard-coded values for x and y that are labeled with the correct values for W and b. Listing 4.9 displays the contents of tf-gradient-descent.py, which illustrates how to compute the MSE with simulated data.

LISTING 4.9: tf-gradient-descent.py

```
import tensorflow as tf

W = tf.Variable([.5], dtype=tf.float32)
b = tf.Variable([-1], dtype=tf.float32)
x = tf.placeholder(tf.float32)
linear_model = W * x + b

y = tf.placeholder(tf.float32)
squared_deltas = tf.square(linear_model - y)
loss = tf.reduce_sum(squared_deltas)

optimizer = tf.train.GradientDescentOptimizer(0.01)
train = optimizer.minimize(loss)

with tf.Session() as sess:
  init = tf.global_variables_initializer()
  sess.run(init) # reset values to incorrect defaults.

  for i in range(1000):
    # Ex1: y = 1*x + 0
    sess.run(train, {x: [1, 2, 3, 4], y: [1, 1, 2, 3]})
    # Ex12 y = 2*x + 0
    #sess.run(train, {x: [1, 2, 3, 4], y: [2, 4, 6, 8]})
    # Ex3: y = 2*x + 1
    #sess.run(train, {x: [1, 2, 3, 4], y: [3, 5, 7, 9]})
    # Ex4: y = -x + 0
    #sess.run(train, {x: [1, 2, 3, 4], y: [-1,-2,-3,-4]})
    # Ex5: y = -x + 1
    #sess.run(train, {x: [1, 2, 3, 4], y: [0,-1,-2,-3]})
    if (i % 50 == 0):
      print(sess.run([W, b]))
```

The first half of Listing 4.9 is very similar to Listing 4.8 (except that W and b have different values). The main difference is in the with code block, which contains five different datasets that correspond to five different lines. For your convenience, the formula for each line is shown above the associated sess. run() statement. Only the first sess.run() statement is executed, data of which is associated with the line y = 1*x + 0.

The final code snippet in Listing 4.9 displays the estimated values for W and b after every 50th iteration of the loop. The output from Listing 4.9 is here:

```
[array([0.82], dtype=float32), array([-0.88], dtype=float32)]
[array([0.8536913], dtype=float32), array([-0.45187122],
                                        dtype=float32)]
[array([0.78414583], dtype=float32), array([-0.24739897],
                                        dtype=float32)]
[array([0.7460697], dtype=float32), array([-0.13545066],
                                        dtype=float32)]
[array([0.7252231], dtype=float32), array([-0.07415909],
                                        dtype=float32)]
[array([0.7138096], dtype=float32), array([-0.040602],
                                        dtype=float32)]
[array([0.7075608], dtype=float32), array([-0.02222957],
                                        dtype=float32)]
[array([0.7041395], dtype=float32), array([-0.01217067],
                                        dtype=float32)]
[array([0.70226634], dtype=float32), array([-0.00666344],
                                        dtype=float32)]
[array([0.70124084], dtype=float32), array([-0.00364819],
                                        dtype=float32)]
[array([0.70067936], dtype=float32), array([-0.00199739],
                                        dtype=float32)]
[array([0.7003719], dtype=float32), array([-0.00109359],
                                        dtype=float32)]
[array([0.70020366], dtype=float32), array([-0.00059871],
                                        dtype=float32)]
[array([0.70011145], dtype=float32), array([-0.00032778],
                                        dtype=float32)]
[array([0.700061], dtype=float32), array([-0.00017944],
                                        dtype=float32)]
[array([0.70003337], dtype=float32), array([-9.823683e-05],
                                        dtype=float32)]
[array([0.7000183], dtype=float32), array([-5.3768184e-05],
                                        dtype=float32)]
[array([0.70001], dtype=float32), array([-2.9393466e-05],
                                        dtype=float32)]
[array([0.7000055], dtype=float32), array([-1.6127855e-05],
                                        dtype=float32)]
[array([0.700003], dtype=float32), array([-8.83344e-06],
                                        dtype=float32)]
```

Now comment out the code snippet for Ex #1, uncomment the code snippet for Ex #2, and launch the code. The new output from Listing 4.9 is here:

```
[array([1.6], dtype=float32), array([-0.62], dtype=float32)]
[array([2.0807753], dtype=float32), array([-0.23748925],
                                        dtype=float32)]
```

```
[array([2.0442245], dtype=float32), array([-0.13002506],
                                            dtype=float32)]
[array([2.0242128], dtype=float32), array([-0.07118864],
                                            dtype=float32)]
[array([2.0132565], dtype=float32), array([-0.0389757],
                                            dtype=float32)]
[array([2.007258], dtype=float32), array([-0.0213393],
                                            dtype=float32)]
[array([2.0039737], dtype=float32), array([-0.01168323],
                                            dtype=float32)]
[array([2.0021756], dtype=float32), array([-0.0063965],
                                            dtype=float32)]
[array([2.0011911], dtype=float32), array([-0.00350218],
                                            dtype=float32)]
[array([2.0006523], dtype=float32), array([-0.00191748],
                                            dtype=float32)]
[array([2.0003572], dtype=float32), array([-0.00104996],
                                            dtype=float32)]
[array([2.0001955], dtype=float32), array([-0.00057487],
                                            dtype=float32)]
[array([2.000107], dtype=float32), array([-0.00031477],
                                            dtype=float32)]
[array([2.0000587], dtype=float32), array([-0.00017239],
                                            dtype=float32)]
[array([2.0000322], dtype=float32), array([-9.457612e-05],
                                            dtype=float32)]
[array([2.0000176], dtype=float32), array([-5.201602e-05],
                                            dtype=float32)]
[array([2.0000095], dtype=float32), array([-2.8360128e-05],
                                            dtype=float32)]
[array([2.0000055], dtype=float32), array([-1.5652413e-05],
                                            dtype=float32)]
[array([2.000003], dtype=float32), array([-8.898016e-06],
                                            dtype=float32)]
[array([2.000002], dtype=float32), array([-5.3002814e-06],
                                            dtype=float32)]
```

Now try the other examples Ex #3 and Ex #4, and then experiment with your own values for the variables x and y. For the sake of simplicity, the data points in all the datasets fit a unique line, which is an oversimplification that enables you easily to compare the correct values with the estimated values. However, later in this chapter you will see datasets whereby the best fitting line does not intersect all the points in the datasets.

CALCULATING MSE WITH NP.LINSPACE() API

The code sample in this section differs from many of the earlier code samples in this chapter: it uses a hard-coded array of values for X, and also for Y, instead of the "perturbation" technique. Hence, you will *not* know the correct value for the slope and y-intercept (and you probably will not be able to guess their correct values). Listing 4.10 displays the contents of plain-linreg1.py that illustrates how to compute the MSE with simulated data.

LISTING 4.10: plain-linreg1.py

```
import numpy as np
import matplotlib.pyplot as plt

X = [0,0.12,0.25,0.27,0.38,0.42,0.44,0.55,0.92,1.0]
Y = [0,0.15,0.54,0.51, 0.34,0.1,0.19,0.53,1.0,0.58]

costs = []
#Step 1: Parameter initialization
W = 0.45
b = 0.75

for i in range(1, 100):
    #Step 2: Calculate Cost
    Y_pred = np.multiply(W, X) + b
    Loss_error = 0.5 * (Y_pred - Y)**2
    cost = np.sum(Loss_error)/10

    #Step 3: Calculate dW and db
    db = np.sum((Y_pred - Y))
    dw = np.dot((Y_pred - Y), X)
    costs.append(cost)

    #Step 4: Update parameters:
    W = W - 0.01*dw
    b = b - 0.01*db

    if i%10 == 0:
        print("Cost at", i,"iteration = ", cost)

#Step 5: Repeat via a for loop with 1000 iterations

#Plot cost versus # of iterations
print("W = ", W,"& b = ",  b)
plt.plot(costs)
plt.ylabel('cost')
plt.xlabel('iterations (per tens)')
plt.show()
```

Listing 4.10 initializes the array variables X and Y with hard-coded values, and then initializes the scalar variables W and b. The next portion of Listing 4.10 contains a for loop that iterates 100 times. After each iteration of the loop, the variables Y_pred, Loss_error, and cost are calculated. Next, the values for dw and db are calculated, based on the sum of the terms in the array Y_pred-Y, and the inner product of Y_pred-y and X, respectively.

Notice how W and b are updated: their values are decremented by the term 0.01*dw and 0.01*db, respectively. This calculation ought to look somewhat familiar: the code is programmatically calculating an approximate value of the gradient for W and b, both of which are multiplied by the learning rate (the hard-coded value 0.01), and the resulting term is decremented from the current values of W and b in order to produce a new approximation for W and b.

Although this technique is very simple, it does calculate reasonable values for w and b.

The final block of code in Listing 4.10 displays the intermediate approximations for w and b, along with a plot of the cost (vertical axis) versus the number of iterations (horizontal axis). The output from Listing 4.10 is here:

```
Cost at 10 iteration =  0.04114630674619492
Cost at 20 iteration =  0.026706242729839392
Cost at 30 iteration =  0.024738889446900423
Cost at 40 iteration =  0.023850565034634254
Cost at 50 iteration =  0.0231499048706651
Cost at 60 iteration =  0.02255361434242207
Cost at 70 iteration =  0.0220425055291673
Cost at 80 iteration =  0.021604128492245713
Cost at 90 iteration =  0.021228111750568435
W =  0.47256473531193927 & b =  0.19578262688662174
```

Figure 4.9 displays a scatter plot of points generated by the code in Listing 4.10.

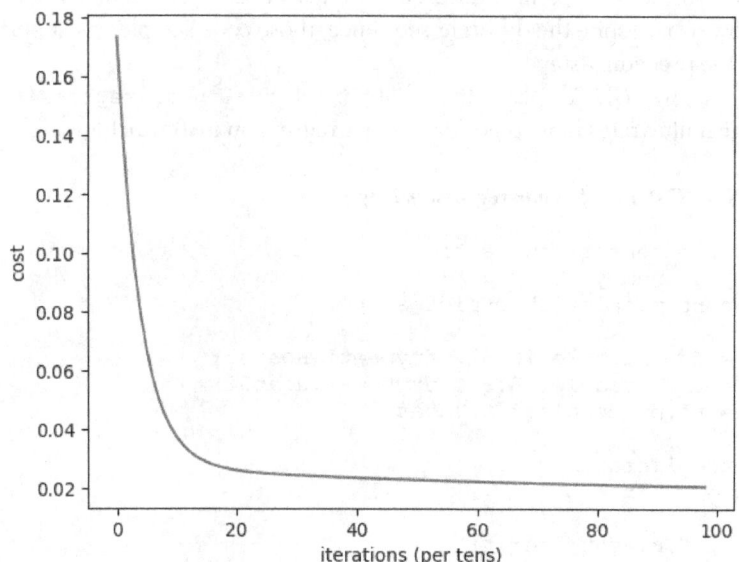

FIGURE 4.9 MSE Values with Linear Regression.

The code sample plain-linreg2.py is similar to the code in Listing 4.10: the difference is that instead of a single loop with 100 iterations, there is an outer loop that execute 100 times, and during each iteration of the outer loop, the inner loop also executes 100 times.

LINEAR REGRESSION WITH TENSORFLOW (1)

The code sample in this section contains primarily TensorFlow code in order to perform linear regression. If you have read the previous examples in this chapter, this section will be easier for you to understand because the steps for linear regression are the same.

Before discussing the code sample, let's look at the general sequence of steps for performing linear regression in TensorFlow:

1. Declare placeholders (X, Y) and variables (W, b)
2. Define the initialization operator (init)
3. Declare operations on placeholders and variables: y_pred, loss, train_op
4. Create a session (sess)
5. Run the initialization operator: sess.run(init)
6. Run some graph operations:

```
sess.run([train_op, loss], feed_dict={X: trainX, Y: trainY})
```

Keep in mind that the preceding sequence of steps is flexible (e.g., steps 1 and 2), and code samples involving TensorFlow and Linear Regression execute the preceding steps in a different order, with different names for variables. However, despite the different sequence, those code samples must perform all of the preceding steps.

Listing 4.11 displays the contents of tf-linear-regression1.py, which illustrates how to perform linear regression in TensorFlow.

LISTING 4.11: tf-linear-regression1.py

```
import tensorflow as tf
import numpy as np
import matplotlib.pyplot as plt

W = tf.Variable([0.5], dtype=tf.float32)
b = tf.Variable([3.6], dtype=tf.float32)
x = tf.placeholder(tf.float32)

# the linear model:
lm = W * x + b

y = tf.placeholder(tf.float32)
sq_deltas = tf.square(lm - y)
loss = tf.reduce_sum(sq_deltas)
lr = 0.001

# specify the following:
# 1) loss function (optimizer)
# 2) learning rate
# 3) minimize loss function
optimizer = tf.train.GradientDescentOptimizer(lr)
train = optimizer.minimize(loss)

# experiment with these values:
factor = 0.25
x_train = np.linspace(-1, 1, 6) # 6 equally spaced numbers
```

```
y_train = 3*x_train+ np.random.randn(*x_train.shape)*factor

init = tf.global_variables_initializer()

# experiment with these values:
iterations = 1000 # 100, 500, 1000
threshold  = 100  # 10,  50,  100

with tf.Session() as sess:
  sess.run(init)

  for i in range(iterations):
    if (i+1) % threshold == 0:
      print(sess.run([W, b]))
    sess.run(train, {x:x_train, y:y_train})

  print('W:',sess.run([W]))
  print('b:',sess.run([b]))

  plt.show()
```

Listing 4.11 defines lm as a linear combination of W, b, and x. The next portion of code defines the loss variable as the MSE of y and 10 variables. Then the code initializes the variable optimizer as a gradient descent optimizer, followed by the train variable that minimizes the loss (which is the MSE).

The next portion of code defines some simulated data by defining the variable x_train as a set of equally spaced numbers between -1 and 1, and the variable y_train as a multiple of the x_train values (plus a small perturbation value), as shown here:

```
factor = 0.25
x_train = np.linspace(-1, 1, 6) # 6 equally spaced numbers
y_train = 3*x_train+ np.random.randn(*x_train.shape)*factor
```

The final block of code contains a loop that trains the model and updates the values of W and b, which are the values of the slope and y-intercept, respectively. The final code snippet displays the computed values for W and b. The output from Listing 4.11 is here:

```
[array([1.5454346], dtype=float32), array([1.0249301],
                                          dtype=float32)]
[array([2.1495297], dtype=float32), array([0.2415321],
                                          dtype=float32)]
[array([2.494051], dtype=float32), array([0.00728351],
                                          dtype=float32)]
[array([2.6905355], dtype=float32), array([-0.06276057],
                                          dtype=float32)]
[array([2.802593], dtype=float32), array([-0.08370487],
                                          dtype=float32)]
[array([2.8665009], dtype=float32), array([-0.08996756],
                                          dtype=float32)]
[array([2.902948], dtype=float32), array([-0.09184019],
                                          dtype=float32)]
```

```
[array([2.9237342], dtype=float32), array([-0.09240016],
                                        dtype=float32)]
[array([2.935589], dtype=float32), array([-0.09256755],
                                        dtype=float32)]
[array([2.9423494], dtype=float32), array([-0.09261762],
                                        dtype=float32)]
W: [array([2.9423997], dtype=float32)]
b: [array([-0.09261787], dtype=float32)]
```

Recall that y_train is defined as three times the x_train values, which means that the slope of the best-fitting line is 3: the calculated value is 2.9423997, which is a decent approximation of the true slope. The value of b (which we didn't know in advance) is calculated as -0.09261787, which is a plausible value.

LINEAR REGRESSION WITH TENSORFLOW (2)

Listing 4.12 displays the contents of tf-linear-regression2.py, which illustrates how to perform linear regression in TensorFlow.

LISTING 4.12: tf-linear-regression2.py

```
from __future__ import print_function

import tensorflow as tf
import numpy
import matplotlib.pyplot as plt
rng = numpy.random

# Hyperparameters
learning_rate = 0.01
training_epochs = 1000
display_step = 50

# Training Data
train_X = numpy.asarray([3.3,4.4,5.5,6.71,6.93,4.168,9.779,
                         6.182,7.59,2.167,
                7.042,10.791,5.313,7.997,5.654,9.
                                        27,3.1])

train_Y = numpy.asarray([1.7,2.76,2.09,3.19,1.694,1.573,
                         3.366,2.596,2.53,1.221,
                2.827,3.465,1.65,2.904,2.42,2.94,
                                        1.3])

n_samples = train_X.shape[0]

# tf Graph Input
X = tf.placeholder("float")
Y = tf.placeholder("float")

# Set model weights
W = tf.Variable(rng.randn(), name="weight")
```

```
b = tf.Variable(rng.randn(), name="bias")

# Construct a linear model
pred = tf.add(tf.multiply(X, W), b)

# Mean squared error
cost = tf.reduce_sum(tf.pow(pred-Y, 2))/(2*n_samples)

# Gradient descent (minimize() modifies W and b)
optimizer = tf.train.GradientDescentOptimizer(learning_
                              rate).minimize(cost)

init = tf.global_variables_initializer()

with tf.Session() as sess:
  # Run the initializer
  sess.run(init)

  # Fit all training data
  for epoch in range(training_epochs):
    for (x, y) in zip(train_X, train_Y):
        sess.run(optimizer, feed_dict={X: x, Y: y})

    # Display logs per epoch step
    if (epoch+1) % display_step == 0:
      c = sess.run(cost, feed_dict={X: train_X, Y:train_Y})
      print("Epoch:",'%04d' % (epoch+1), "cost=", "{:.9f}".
                                    format(c), \
            "W=", sess.run(W), "b=", sess.run(b))

  print("Optimization completed.")
  training_cost = sess.run(cost, feed_dict={X: train_X, Y:
                                    train_Y})
  print("Training cost=",training_cost,"W=",sess.
                      run(W),"b=",sess.run(b),'\n')

  # display graph
  plt.plot(train_X, train_Y, 'ro', label='Original data')
  plt.plot(train_X, sess.run(W)*train_X+sess.run(b),
                              label='Fitted line')
  plt.legend()
  plt.show()

  # Testing example
  test_X = numpy.asarray([6.83, 4.668, 8.9, 7.91, 5.7, 8.7,
                              3.1, 2.1])
  test_Y = numpy.asarray([1.84, 2.273, 3.2, 2.831, 2.92,
                              3.24, 1.35, 1.03])

  print("Working with test-related data")
  testing_cost = sess.run(
      tf.reduce_sum(tf.pow(pred - Y, 2)) / (2 * test_X.
                              shape[0]),
      feed_dict={X: test_X, Y: test_Y})  # same function as
                              cost above
```

```
print("Testing cost=", testing_cost)
print("Absolute mean square loss difference:", abs(
    training_cost - testing_cost))

plt.plot(test_X, test_Y, 'bo', label='Testing data')
plt.plot(train_X, sess.run(W) * train_X+sess.run(b),
                                    label='Fitted line')
plt.legend()
plt.show()
```

Listing 4.12 starts by initializing the variables learning_rate and train-ing_epochs, both of which are hyper parameters, followed by the variable display_step that is used to determine when to display intermediate results.

The next portion of Listing 4.12 uses the np.array() method in order to generate data for the array variables train_X and train_Y, followed by n_samples (the number of elements in the array train_X), TF placehold-ers X and Y, and the variable W (the slope of the line) and the variable b (the y-intercept).

Next, the TF variable pred (for "predicted value") is initialized as a linear combination of W and b. The TF variables cost, optimizer, and init are initialized in the same manner that you have seen in previous code samples.

The second half of Listing 4.12 consists of the with code block, which contains a nested loop. The outer loop iterates through the number of epochs, and the inner loop trains the linear model by iterating through the values of train_X and train_Y via the TF placeholders X and Y, respectively. The output from Listing 4.12 is here:

```
Epoch: 0050 cost= 0.108249746 W= 0.348563 b= 0.089487664
Epoch: 0100 cost= 0.104633830 W= 0.34267682 b= 0.13183235
Epoch: 0150 cost= 0.101435706 W= 0.33714065 b= 0.1716589
Epoch: 0200 cost= 0.098607130 W= 0.33193392 b= 0.20911649
Epoch: 0250 cost= 0.096105486 W= 0.32703662 b= 0.24434637
Epoch: 0300 cost= 0.093892939 W= 0.32243082 b= 0.27748084
Epoch: 0350 cost= 0.091936104 W= 0.3180987 b= 0.3086453
Epoch: 0400 cost= 0.090205535 W= 0.31402448 b= 0.33795545
Epoch: 0450 cost= 0.088675007 W= 0.3101924 b= 0.36552286
Epoch: 0500 cost= 0.087321416 W= 0.30658814 b= 0.3914513
Epoch: 0550 cost= 0.086124368 W= 0.30319846 b= 0.4158368
Epoch: 0600 cost= 0.085065775 W= 0.3000103 b= 0.4387719
Epoch: 0650 cost= 0.084129564 W= 0.29701185 b= 0.46034324
Epoch: 0700 cost= 0.083301708 W= 0.2941915 b= 0.48063222
Epoch: 0750 cost= 0.082569622 W= 0.29153916 b= 0.49971336
Epoch: 0800 cost= 0.081922255 W= 0.28904438 b= 0.5176604
Epoch: 0850 cost= 0.081349798 W= 0.2866981 b= 0.5345395
Epoch: 0900 cost= 0.080843613 W= 0.2844912 b= 0.5504155
Epoch: 0950 cost= 0.080396011 W= 0.28241557 b= 0.56534725
Epoch: 1000 cost= 0.080000252 W= 0.28046343 b= 0.57939106
Optimization Completed.
Training cost= 0.08000025 W= 0.28046343 b= 0.57939106

Working with test-related data
Testing cost= 0.07633824
Absolute mean square loss difference: 0.0036620125
```

Figure 4.10 displays a scatter plot of points (for the training portion of the dataset) generated by the code in Listing 4.12.

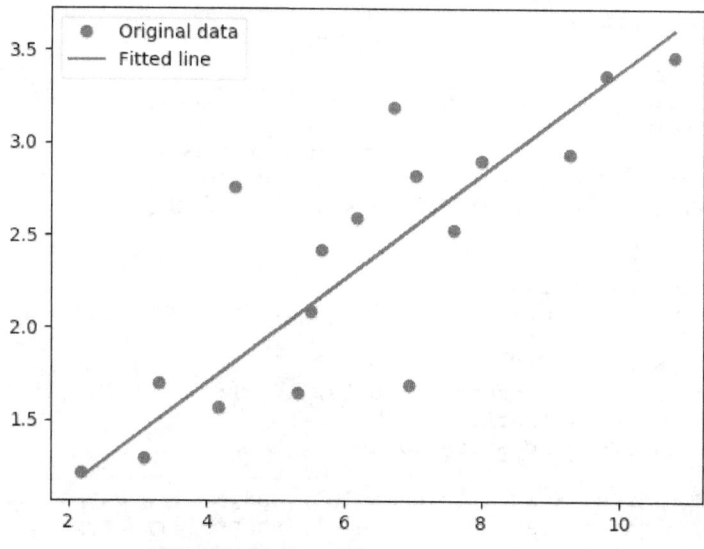

FIGURE 4.10 Linear Regression in TensorFlow.

The companion disc contains `simple-linreg2.py` and `simple-adam-linreg1.py`, which involve linear regression and the Adam gradient descent.

WORKING WITH TF.ESTIMATOR MODELS

When you create a `tf.estimator` model, you specify the input data via an "input function" called `input_fn`. This function, which returns a `tf.data.Dataset` of batches of data pairs, is invoked from `tf.estimator.Estimator` methods such as "train" and "evaluate." The input function returns the following pair of data values:

- features: A dict from feature names to Tensors or SparseTensors with batches of features
- labels: A Tensor containing batches of labels

The keys of the "features" are used to configure the model's input layer.

Note: The input function is called while constructing the TensorFlow graph, °not° while running the graph. It is returning a representation of the input data as a sequence of TensorFlow graph operations.

LINEAR REGRESSION WITH TF.ESTIMATOR

Listing 4.13 displays the contents of `tf-estimator-linear.py`, which illustrates how to calculate the value of `m` (the slope) and `b` (the y-intercept) of the best fitting line for a set of points in the plane.

LISTING 4.13: tf-estimator-linear.py

```
import tensorflow as tf
import numpy as np

factor = 1.0 # 0.25
x_train = np.linspace(1, 4, 4) # 4 equally spaced numbers
y_train = 3*x_train
y_train = 3*x_train + np.random.randn(*x_train.shape)*factor
x_eval  = np.array([2., 5., 8., 10.])
y_eval  = np.array([6.01, 14.1, 25, 32.])


# prediction inputs
x_predict = np.array([25.0, 50.0, 75.0, 100.0])

# linear model: y = Wx + b
def model_fn(features, labels, mode):
  if mode == 'infer':
    labels = np.array([0, 0])

  W = tf.get_variable(name='W', shape=[1], dtype=tf.float64)
  b = tf.get_variable(name='b', shape=[1], dtype=tf.float64)
  y = W * features['x'] + b

  loss = tf.reduce_sum(input_tensor=tf.square(x=(y -
                                          labels)))
  optimizer = tf.train.GradientDescentOptimizer(learning_
                                          rate=0.01)
  global_step = tf.train.get_global_step()

  train_step = tf.group(optimizer.minimize(loss=loss),
                        tf.assign_add(global_step, 1))

  return tf.estimator.EstimatorSpec(
      mode=mode,
      predictions=y,
      loss=loss,
      train_op=train_step
)

# model inputs: we need another feature column for
# each feature/factor that we specify as input
feature_column = tf.feature_column.numeric_column(key='x',
                                          shape=[1])
feature_columns = [feature_column]

# Type of function to use (ex: Linear Regression)
estimator = tf.estimator.Estimator(model_fn=model_fn)

# Input function for training the model using training data
input_fn = tf.estimator.inputs.numpy_input_
                              fn(x={'x': x_train},
                                 y=y_train,
                                 batch_size=4,
                                 num_epochs=None,
                                 shuffle=True)
```

```
# Input function to evaluate the training phase of the model
train_input_fn = tf.estimator.inputs.numpy_input_
                                fn(x={'x': x_train},
                                   y=y_train,
                                   batch_size=4,
                                   num_epochs=1000,
                                   shuffle=False)

# Input function to test the model
eval_input_fn = tf.estimator.inputs.numpy_input_
                                fn(x={'x': x_eval},
                                   y=y_eval,
                                   batch_size=4,
                                   num_epochs=1000,
                                   shuffle=False)

# Input function to predict values
predict_input_fn = tf.estimator.inputs.numpy_input_
                                fn(x={'x': x_predict},
                                   num_epochs=1,
                                   shuffle=False)

# training phase: 1000 epochs and the training function
estimator.train(input_fn=input_fn, steps=1000)

# evaluate the training phase
print("train:    ",estimator.evaluate(input_fn=train_input_
                                                        fn))

# evaluate the testing phase (final performance)
print("testing:  ",estimator.evaluate(input_fn=eval_input_
                                                        fn))

# predict using x_predict values and trained model
print("x_predict:",x_predict)
print("predicted:",list(estimator.predict(input_fn=predict_
                                                input_fn)))

# input_fn, train_input_fn, and eval_input_fn require:
# batch_size, num_epochs, and shuffle values

input_fn = tf.estimator.inputs.numpy_input_fn(
    {"x": x_train}, y_train, batch_size=4, num_epochs=None,
                                            shuffle=True)

train_input_fn = tf.estimator.inputs.numpy_input_fn(
    {"x": x_train}, y_train, batch_size=4, num_epochs=1000,
                                            shuffle=False)

eval_input_fn = tf.estimator.inputs.numpy_input_fn(
    {"x": x_eval}, y_eval, batch_size=4, num_epochs=1000,
                                            shuffle=False)

# 'steps' = # of training steps:
estimator.train(input_fn=input_fn, steps=1000)
```

```
# evaluate the accuracy of our model:
train_metrics = estimator.evaluate(input_fn=train_input_fn)
eval_metrics = estimator.evaluate(input_fn=eval_input_fn)
print("train metrics: %r"% train_metrics)
print("eval metrics: %r"% eval_metrics)
```

Listing 4.13 initializes x_train and y_train via the np.linspace() API in order to define the line y = 3*x + (perturbation factor).

The next portion of Listing 4.13 defines the Python function model_fn that defines a linear regression model, the loss function, and the optimizer. The return value is an instance of the tf.estimator.EstimatorSpec class that includes values for mode, predictions, loss, and train_op.

The next portion of Listing 4.13 defines feature_columns as an array of columns that happens to be just the value "x" (in general this array can contain many values). The middle portion of Listing 4.13 defines the four input functions input_fn, train_input_fn, eval_input_fn, and predict_input_fn as the functions for the input data, for training phase, for the testing phase, and for the prediction phases, respectively. All of these functions require specifying suitable values for x, y, batch_size, num_epochs, and shuffle.

The next code block in Listing 4.13 performs the actual training of the model for 1,000 epochs, then evaluates the training phase as well as the testing phase, and lastly makes predictions. The output from launching the code in Listing 4.13 is here:

```
('train:     ', {'loss': 2.0612717, 'global_step': 1000})
('testing:   ', {'loss': 24.034492, 'global_step': 1000})
('x_predict:', array([ 25.,   50.,   75., 100.]))
('predicted:', [69.19919776290799, 138.07798532975252,
               206.95677289659704, 275.83556046344154])
```

As you can see, the predicted values are approximate three times the corresponding values in the x_predict array. This shows you that a small dataset, such as the one in Listing 4.13, is unlikely to result in highly accurate predictions.

SUMMARY

This chapter introduced you to linear regression, and a brief description of how to calculate a best-fitting line for a dataset of values in the Euclidean plane. You saw how to perform linear regression using NumPy in order to initialize arrays with data values, along with a "perturbation" technique that introduces some randomness for the y values. This technique is useful because you will know the correct values for the slope and y-intercept of the best-fitting line, which you can then compare with the trained values.

You then learned how to perform linear regression in code samples that involve TensorFlow. In addition, you saw how to use Matplotlib in order to display line graphs for best-fitting lines and graphs that display the cost versus the number of iterations during the training-related code blocks. Finally, you saw how to use a tf.estimator class in order to train a linear model.

CHAPTER 5

LOGISTIC REGRESSION

This chapter assumes that you have been exposed to activation functions (even if you do not fully understand them), and also that you have read the material in Chapter 4. In addition, you need a basic understanding of hidden layers in a neural network, which is not discussed in this book. Depending on your comfort level, you might need to read some preparatory material before diving into this chapter (there are many articles available online).

The rationale for the preceding assumptions is that this book is not for "absolute beginners" (as mentioned in the Preface), and providing a very detailed explanation for every concept in every code samples in every chapter would probably double the length of this book. However, this book belongs to the Pocket Primer series of books for advanced beginners, which assumes that readers have already been exposed (to a limited extent) to some of the concepts in this chapter.

As such, this book attempts a compromise: explanations are included when they can extend your current knowledge and at the same time also "cut to the chase" (in a manner of speaking) regarding the code. Depending on your background, be prepared to spend some time reading the code samples, and feel free to "tweak" the values of parameters to see how they affect the performance of the code.

As you will see, this chapter contains fewer code samples that are somewhat lengthier than the code samples in the previous chapter. In some cases there is a list of TF APIs in a comment block near the top of the code samples. Some code samples are divided into subsections to make the code a bit easier to grasp. Don't be surprised if you need to read the code samples more than once, and more importantly, don't be discouraged by the learning curve. As a sort of consolation: after you've worked with TF 1.x you will discover that TF 2 is easier than TF 1.x and your learning curve will probably be significantly shorter.

With the preceding points in mind, the first section of this chapter briefly discusses linear classifiers, followed by an overview of activation functions, which are of paramount importance for deep neural networks. In this section, you will learn how and why they are used in neural networks. This section also contains a list of the TensorFlow APIs for activation functions, followed by a description of some of their merits. Even though Logistic Regression relies on the sigmoid function, it's still worthwhile to learn something about some of the other activation functions. In addition, you will see a code sample that contains TensorFlow and the sigmoid activation function.

The second section introduces logistic regression, which is actually a classifier (despite its name). This section contains a code sample that involves TensorFlow and Logistic Regression. The final portion of this chapter contains a very brief introduction to Keras, along with a code sample that illustrates how to define a Keras-based model for the MNIST dataset. Keras is well-integrated into the latest versions of TF 1.x, and it's also a core part of TensorFlow 2. Later on, if you plan to learn TensorFlow 2, it's definitely worthwhile to learn more about Keras.

WHAT IS CLASSIFICATION?

The purpose of classification is to predict the class of given data points, where classes refer to categories and are also called targets or labels. For example, spam detection in email service providers involves binary classification (only 2 classes). The MNIST dataset contains a set of images, where each image is a single digit, which means there are 10 labels. Some applications in classification include: credit approval, medical diagnosis, and target marketing.

What are Classifiers?

In the previous chapter you learned that linear regression uses supervised learning in conjunction with numeric data: the goal is to train a model that can make numeric predictions (e.g., the price of stock tomorrow, the temperature of a system, its barometric pressure, and so forth).

By contrast, classifiers use supervised learning in conjunction with non-numeric classes of data: the goal is to train a model that can make categorical predictions. For instance, suppose that each row in a dataset is a specific wine, and each column pertains to a specific wine feature (tannin, acidity, and so forth).

Suppose further that there are five classes of wine in the dataset: for simplicity, let's label them A, B, C, D, and E. Given a new data point, which is to say a new row of data, a classifier for this dataset attempts to determine the label for this new wine.

Some of the classifiers in this chapter can perform categorical classification and also make numeric predictions (i.e., they can be used for regression as well as classification).

Common Classifiers

Some of the most popular classifiers for Machine Learning are listed here:

- linear classifiers
- kNN
- logistic regression
- decision trees
- random forests
- SVMs
- Bayesian classifiers
- CNNs (deep learning)

In the case of Deep Learning, image classification can be performed by CNNs (Convolutional Neural Networks), which makes them classifiers (they can also be used for audio and text processing).

Keep in mind that different classifiers have different advantages and disadvantages, which often involves a trade-off between complexity and accuracy, similar to algorithms in fields that are outside of AI.

The subsequent sections provide a brief description of the ML classifiers that are listed in the previous section.

What are Linear Classifiers?

A linear classifier separates a dataset into two classes. A linear classifier is a line for 2D points, a plane for 3D points, and a hyper plane (a generalization of a plane) for higher dimensional points.

Linear classifiers are often the fastest classifiers, so they are often used when the speed of classification is of high importance. Linear classifiers usually work well when the input vectors are sparse (i.e., mostly zero values) or when the number of dimensions is large.

WHAT ARE ACTIVATION FUNCTIONS?

The briefest possible description: an activation function is (usually) a nonlinear function that introduces non-linearity into a neural network, thereby preventing a "consolidation" of the hidden layers in neural network. Some well-known activation functions are sigmoid, tanh, and ReLU (which also has several variations).

Suppose that we have a neural network that does not contain any activation function. The weights of the edges that connect the input layer with the first hidden layer is a matrix: let's call it W1. Next, the weights of the edges that connect the first hidden layer with the second hidden layer is a matrix: let's call it W2. Repeat this process until we reach the edges that connect the final hidden layer with the output layer: let's call this matrix Wk. Since we do not have an activation function, we can simply multiply the matrices W1, W2, ...,

Wk together and produce *one* matrix: let's call it W. We have now collapsed the original neural network with one involving an input layer, a matrix of weights W, and an output layer. In other words, we no longer have a neural network!

Fortunately, this reduction will not occur when we specify an activation function between every pair of consecutive layers. In other words, *an activation function at each layer prevents this "matrix consolidation."* Hence, we can maintain all the intermediate hidden layers during the process of training the neural network.

For simplicity, let's assume that we have the same activation function between every pair of adjacent layers (we'll remove this assumption shortly). The process for using an activation function in a neural network is as follows:

1. start with an input vector x1 of numbers
2. multiply x1 by the matrix of weights W1 for the edges that connect the input layer with the first hidden layer, which produces a new vector x2
3. "apply" the activation function to each element of x2 to create another vector x3

At this point we repeat Steps 2 and 3, except that we use the "starting" vector x3 and the weights matrix W2 for the edges that connect the first hidden layer with the second hidden layer (or just the output layer if there is only one hidden layer).

After completing the preceding process, we have "preserved" the neural network, which means that it can be trained on a dataset. One other thing: instead of using the same activation function at each step, you can replace each activation function by a different activation function (the choice is yours).

Why do we Need Activation Functions?

The previous section outlines the process for transforming an input vector from the input layer and then through the hidden layers until it reaches the output layer. The purpose of activation functions in neural networks is vitally important, so it's worth repeating here: activation functions "preserve" the structure of neural networks and prevent them from being reduced to an input layer and an output layer. In other words, if we specify a non-linear activation function between every pair of consecutive layers, then we explicitly control the exact number of layers in the neural network.

Without a non-linear activation function, we simply multiply a weight matrix for a given pair of consecutive layers with the output vector that is produced from the previous pair of consecutive layers. We repeat this simple multiplication until we reach the output layer of the neural network.

How do Activation Functions Work?

If this is the first time you have encountered the concept of an activation function, it's probably confusing, so here's an analogy that might be helpful.

Suppose you're driving your car late at night and there's nobody else on the highway. You can drive at a constant speed for as long as there are no obstacles (stop signs, traffic lights, and so forth). On the other hand, suppose you drive into the parking lot of a large grocery store. When you approach a speed bump you must slow down, cross the speed bump, and increase speed again, and repeat this process for every speed bump.

Think of the non-linear activation functions in a neural network as the counterpart to the speed bumps: you simply cannot maintain a constant speed, which (by analogy) means that you cannot first multiply all the weight matrices together and "collapse" them into a single weight matrix. Another analogy involves a road with multiple toll booths: you must slow down, pay the toll, and then resume driving until you reach the next toll booth. These are only analogies (and hence imperfect) to help you understand the need for non-linear activation functions.

Common Activation Functions

Although there are many activation functions (and you can define your own if you know how to do so), here is a list of common activation functions, followed by brief descriptions:

- Sigmoid
- Tanh
- ReLU
- ReLU6
- ELU
- SELU

The `sigmoid` activation function is based on Euler's constant e, with a range of values between 0 and 1, and its formula is shown here:

```
1/[1+e^(-x)]
```

The `tanh` activation function is also based on Euler's constant e, and its formula is shown here:

```
[e^x - e^(-x)]/[e^x+e^(-x)]
```

One way to remember the preceding formula is to note that the numerator and denominator have the same pair of terms: they are separated by a "-" sign in the numerator and a "+" sign in the denominator. The `tanh` function has a range of values between -1 and 1.

The ReLU (Rectified Linear Unit) activation function is straightforward: if x is negative then ReLU(x) is 0; for all other values of x, ReLU(x) equals x. ReLU6 is specific to TensorFlow, and it's a variation of ReLU(x): the additional constraint is that ReLU(x) equals 6 when x >= 6 (hence its name).

ELU is Exponential Linear Unit and it's the exponential "envelope" of ReLU, which replaces the two linear segments of ReLU with an exponential activation function that is differentiable for all values of x (including x = 0).

SELU is an acronym for Scaled Exponential Linear Unit, and it's slightly more complicated than the other activation functions (and used less frequently). For a thorough explanation of these and other activation functions (along with graphs that depict their shape), navigate to the following Wikipedia link:

https://en.wikipedia.org/wiki/Activation_function

The preceding link provides a long list of activation functions as well as their derivatives.

TENSORFLOW ACTIVATION FUNCTIONS

TensorFlow (and many other frameworks) provide implementations for many activation functions, which saves you the time and effort from writing your own implementation of activation functions.

Here is a list of TensorFlow APIs that correspond to some of the activation functions that are listed in the previous section:

- tf.nn.relu
- tf.nn.relu6
- tf.nn.selu
- tf.nn.softmax
- tf.nn.softmax_cross_entropy_with_logits_v2
- tf.nn.softplus
- tf.sigmoid
- tf.tanh

The following subsections provide additional information regarding some of the activation functions in the preceding list. Keep the following point in mind: for simple neural networks, use ReLU as your first preference.

The ReLU Activation Function

Currently ReLU is often the "preferred" activation function: previously the preferred activation function was `tanh` (and before `tanh` it was `sigmoid`). ReLU behaves close to a linear unit and provides the best training accuracy and validation accuracy.

ReLU is like a switch for linearity: it's "off" if you don't need it, and its derivative is 1 when it's active, which makes ReLU the simplest of all the current activation functions. The second derivative of the function is 0 (except the origin): it's a very simple function that simplifies optimization. In addition, the gradient is large whenever you need large values, and it never "saturates" (i.e., it does not shrink to zero on the positive horizontal axis).

Rectified linear units and generalized versions are based on the principle that linear models are easier to optimize. Use the ReLU activation function or one of its related alternatives (discussed later).

The Advantages and Disadvantages of ReLU

The following list contains the advantages of the ReLU activation function:

- Does not saturate in the positive region
- very efficient in terms of computation
- models with ReLU typically converge faster those with other activation functions

However, ReLU does have a disadvantage when the activation value of a ReLU neuron becomes 0: then the gradients of the neuron will also be 0 during back-propagation. You can mitigate this scenario by judiciously assigning the values for the initial weights as well as the learning rate.

ELU

ELU is an acronym for *exponential linear unit* that is based on ReLU: the key difference is that ELU is differentiable at the origin (ReLU is a continuous function but *not* differentiable at the origin). However, keep in mind several points. First, ELUs trade computational efficiency for "immortality" (immunity to dying): read the following paper for more details: arxiv.org/abs/1511.07289. Secondly, RELUs are still popular and preferred over ELU because the use of ELU introduces an additional new hyper-parameter.

Sigmoids

The `sigmoid` activation function has a range in (0,1), and it saturates and "kills" gradients. Unlike the `tanh` activation function, `sigmoid` outputs are not zero-centered. In addition, both `sigmoid` and `softmax` (discussed later) are discouraged for vanilla feed forward implementation (see Chapter 6 of *Deep Learning*, the online book by Ian Goodfellow et al.). However, the `sigmoid` activation function is still used in LSTMs (specifically for the forget gate, input gate, and the output gate), GRUs (Gated Recurrent Units), and probabilistic models. Moreover, some autoencoders have additional requirements that preclude the use of piecewise linear activation functions.

Softmax

The `softmax` activation function maps the values in a dataset to another set of values that are between 0 and 1, and whose sum equals 1. Thus, `softmax` creates a probability distribution. In the case of image classification with Convolutional Neural Networks (CNNs), the `softmax` activation function "maps" the values in the final hidden layer (often abbreviated as "FC") to the 10 neurons in the output layer. The index of the position that contains the largest probability is matched with the index of the number 1 in the one-hot encoding of the input image. If the index values are equal, then the image has been classified, otherwise it's considered a mismatch.

Softplus

The `softplus` activation function is a smooth (i.e., differentiable) approximation to the ReLU activation function. Recall that the origin is the only non-differentiable point of the ReLU function, which is "smoothed" by the `softmax` activation whose equation is here:

```
f(x) = ln(1 + e^x)
```

Tanh

The `tanh` activation function has a range in (-1,1), whereas the `sigmoid` function has a range in (0,1). Both of these two activations saturate, but unlike the `sigmoid` neuron the `tanh` output is zero-centered. Therefore, in practice the `tanh` non-linearity is always preferred to the `sigmoid` nonlinearity.

Although `sigmoid` and `tanh` activation functions are not the preferred activation functions for Convolutional Neural Networks, both of these activation functions appear in LSTMs (and other RNNs) during the calculations pertaining to input gates, forget gates, and output gates (all of which is beyond the scope of this book).

SIGMOID, SOFTMAX, AND HARDMAX

This section briefly discusses some of the differences among these three functions. First, the `sigmoid` function is used for binary classification in logistic regression model, as well as the gates in LSTMs and GRUs. The `sigmoid` function is used as activation function while building neural networks, but keep in mind that the sum of the probabilities is *not* necessarily equal to 1.

Second, the `softmax` function generalizes the `sigmoid` function: it's used for multi-classification in logistic regression model. The `softmax` function is the activation function for the "fully connected layer" in CNNs, which is the right-most hidden layer and the output layer. Unlike the sigmoid function, the sum of the probabilities *must* equal 1. You can use either the sigmoid function or `softmax` for binary (n=2) classification.

Third, the so-called "`hardmax`" function assigns 0 or 1 to output values (similar to a step function). For example, suppose that we have three classes {c1, c2, c3} whose scores are [1, 7, 2], respectively. The `hardmax` probabilities are [0, 1, 0], whereas the `softmax` probabilities are [0.1, 0.7, 0.2]. Notice that the sum of the `hardmax` probabilities is 1, which is also true of the sum of the `softmax` probabilities. However, the `hardmax` probabilities are all-or-nothing, whereas the `softmax` probabilities are analogous to receiving "partial credit."

TENSORFLOW AND THE SIGMOID ACTIVATION FUNCTION

Listing 5.1 displays the contents of `tf-activation-functions.py` that illustrates how to create a TensorFlow graph that involves seven activation functions, including the `sigmoid` function for logistic regression.

LISTING 5.1: tf-activation-functions.py

```
import matplotlib.pyplot as plt
import numpy as np
import tensorflow as tf
from tensorflow.python.framework import ops
ops.reset_default_graph()

sess = tf.Session()

# X range
x_vals = np.linspace(start=-10., stop=10., num=100)

# ReLU activation
print(sess.run(tf.nn.relu([-3., 3., 10.])))
y_relu = sess.run(tf.nn.relu(x_vals))

# ReLU-6 activation
print(sess.run(tf.nn.relu6([-3., 3., 10.])))
y_relu6 = sess.run(tf.nn.relu6(x_vals))

# Sigmoid activation
print(sess.run(tf.nn.sigmoid([-1., 0., 1.])))
y_sigmoid = sess.run(tf.nn.sigmoid(x_vals))

# Hyperbolic Tangent activation
print(sess.run(tf.nn.tanh([-1., 0., 1.])))
y_tanh = sess.run(tf.nn.tanh(x_vals))

# Softsign activation
print(sess.run(tf.nn.softsign([-1., 0., 1.])))
y_softsign = sess.run(tf.nn.softsign(x_vals))

# Softplus activation
print(sess.run(tf.nn.softplus([-1., 0., 1.])))
y_softplus = sess.run(tf.nn.softplus(x_vals))

# Exponential linear activation (ELU)
print(sess.run(tf.nn.elu([-1., 0., 1.])))
y_elu = sess.run(tf.nn.elu(x_vals))

# Plot the different functions
plt.plot(x_vals, y_softplus, 'r--', label='Softplus',
                                     linewidth=2)
plt.plot(x_vals, y_relu, 'b:', label='ReLU', linewidth=2)
plt.plot(x_vals, y_relu6, 'g-.', label='ReLU6',
                                     linewidth=2)
```

```
plt.plot(x_vals, y_elu, 'k-', label='ExpLU', linewidth=0.5)
plt.ylim([-1.5,7])
plt.legend(loc='best')
plt.show()

plt.plot(x_vals, y_sigmoid, 'r--', label='Sigmoid',
                                            linewidth=2)
plt.plot(x_vals, y_tanh, 'b:', label='Tanh', linewidth=2)
plt.plot(x_vals, y_softsign, 'g-.', label='Softsign',
                                            linewidth=2)
plt.ylim([-2,2])
plt.legend(loc='best')
plt.show()
```

Listing 5.1 starts with some `import` statements, followed by the `sess` variable that is an instance of the `tf.Session()` class. The next code snippet initializes the `x_vals` array, which contains 100 equally spaced numbers in the interval (-10, 10).

The next block of extensive code in Listing 5.1 shows you how to invoke the TensorFlow activation functions that are listed in a previous section. The final section of code in Listing 5.1 plots the various TensorFlow functions.

Figure 5.1 displays the graph of the TensorFlow activation functions that are defined in the first portion of Listing 5.1.

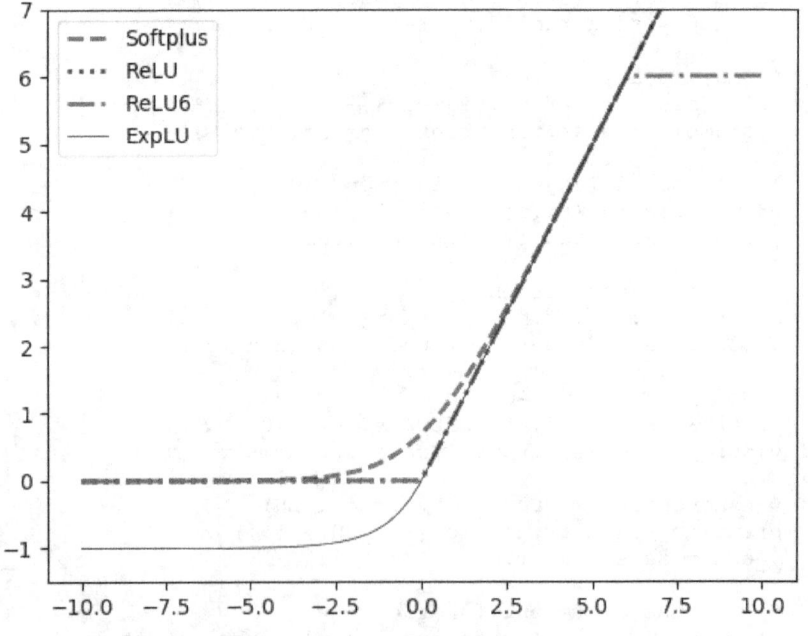

FIGURE 5.1 TensorFlow Activation Functions.

Figure 5.2 displays the graph of the TensorFlow activation functions that are defined in the second portion of Listing 5.1.

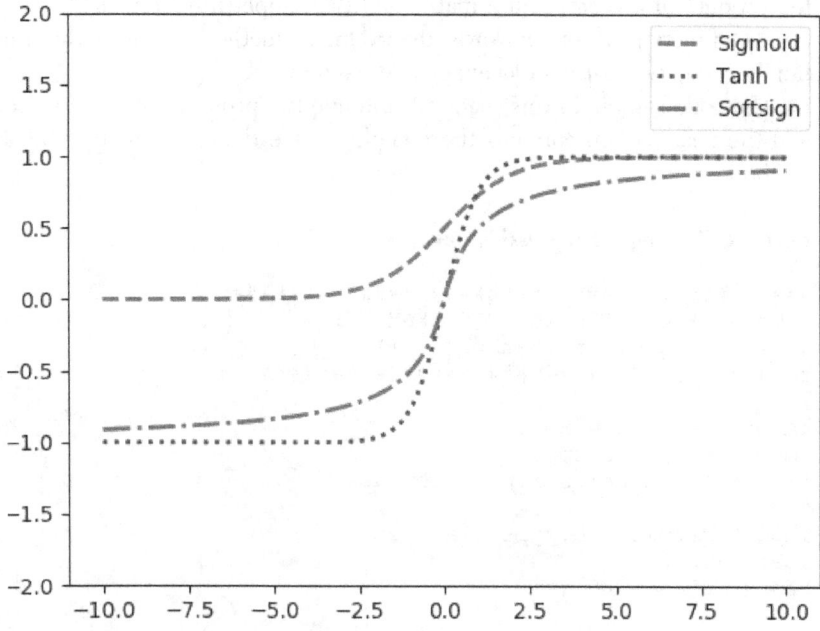

FIGURE 5.2 TensorFlow Activation Functions.

WHAT IS LOGISTIC REGRESSION?

Logistic regression is essentially the result of "applying" the `sigmoid` activation function to linear regression in order to perform binary classification. Logistic regression is useful in a variety of unrelated fields. Such fields include machine learning, various medical fields, and social sciences. Logistic regression can be used to predict the risk of developing a given disease, based on various observed characteristics of the patient. Other fields that use logistic regression include engineering, marketing, and economics.

Logistic regression can be binomial (only two outcomes for a dependent variable), multinomial (three or more outcomes for a dependent variable), or ordinal (dependent variables are ordered).

For instance, suppose that a dataset consists of data that belong either to class A or to class B. If you are given a new data point, logistic regression predicts whether that new data point belongs to class A or to class B. By contrast, linear regression predicts a numeric value, such as the next-day value of a stock.

Despite its name, logistic regression is a classification technique and not a regression technique.

TENSORFLOW AND LOGISTIC REGRESSION (1)

Listing 5.2 displays the contents of `logistic-regression1.py` that defines (but does not invoke) the Python method `model()` that calculates the product of a vector and a matrix and then applies the TensorFlow `sigmoid()` function. In other words, the `model()` method performs the transition between two adjacent layers in a neural network.

The code sample in this section is intended to present the "set-up" code, and the next section contains the complete code that invokes the `model()` method.

LISTING 5.2: logistic-regression1.py

```
###########################################
# TensorFlow APIs IN THIS EXAMPLE:
# tf.nn.sigmoid(tf.matmul(...))
###########################################

import tensorflow as tf
import numpy as np
from sklearn.utils import shuffle

import matplotlib.pyplot as plt
trainsamples = 200
testsamples = 60

# the model, a simple input, a hidden layer (sigmoid
                                             activation)
def model(X, hidden_weights1, hidden_bias1, ow):
  hidden_layer =  tf.nn.sigmoid(tf.matmul(X, hidden_
                                             weights1)+ b)
  return tf.matmul(hidden_layer, ow)

dsX = np.linspace(-1, 1, trainsamples + testsamples).
                                             transpose()
dsY = 0.4*pow(dsX,2)+2*dsX+np.random.randn(*dsX.
                                             shape)*0.22+0.8

plt.figure()
plt.title('Original data')
plt.scatter(dsX,dsY)
plt.show()
```

Listing 5.2 starts with some `import` statements, including one for the SKLearn `shuffle` API that is used in the complete code sample in the next section. The next portion of Listing 5.2 initializes the variables `trainsamples` and `testsamples` with the number of training samples and test samples, respectively.

The next portion of Listing 5.2 defines the `model()` method, which takes the parameters X, `hidden_weights`, `hidden_bias1`, and ow. The `model()` method has just two lines of code. The first line of code computes the `hidden_layer` vector by invoking the TensorFlow `tf.nn.sigmoid()` method to the result of multiplying the input vector X with the `hidden_weights1` matrix (and then adding the bias vector b). The second line of code multiplies `hidden_layer` and ow, and then returns the resulting product.

The next section of Listing 5.2 initializes the Python array variables `dsX` and `dsY`. The variable `dsX` consists of an array of evenly numbers of size `(trainsamples+testsamples)` in the interval (-1,1). On the other hand, the variable `dsY` consists of an array of numbers that involve a quadratic equation. The first term in the quadratic equation is `0.4*pow(dsX,2)`, and the second term is a product of a random number and a number in the `dsX` array (as well as another constant), and the third term is the constant 0.8. The two definitions are reproduced here:

```
dsX = np.linspace(-1, 1, trainsamples + testsamples).
                                          transpose()
dsY = 0.4*pow(dsX,2)+2*dsX+np.random.randn(*dsX.
                                 shape)*0.22+0.8
```

The final section of Listing 5.2 is some plot-related code that is already familiar to you from many previous examples.

Figure 5.3 displays the output from launching the code in Listing 5.2.

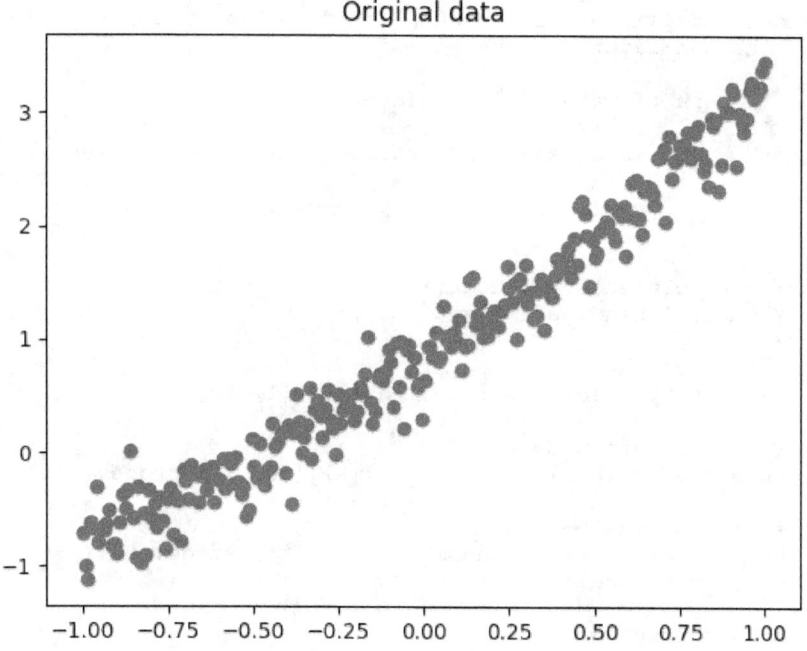

FIGURE 5.3 A ScatterPlot of Quadratic Points.

TENSORFLOW AND LOGISTIC REGRESSION (2)

Listing 5.3 displays the contents of `logistic-regression2.py` that illustrates how to create a TensorFlow graph that uses logistic regression. Note that this code sample extends the code in Listing 5.2.

LISTING 5.3: logistic-regression2.py

```python
#############################################
# TensorFlow APIs IN THIS EXAMPLE:
# tf.nn.sigmoid(tf.matmul(...))
# tf.Variable(tf.random_normal(...))
# tf.placeholder(...)
# tf.pow(model_y-Y, 2)/(2)
# tf.global_variables_initializer().run()
#############################################

import tensorflow as tf
import numpy as np
from sklearn.utils import shuffle

import matplotlib.pyplot as plt
trainsamples = 200
testsamples = 60

#the model, a simple input, a hidden layer (sigmoid
                                          activation)
def model(X, hidden_weights1, hidden_bias1, ow):
  hidden_layer = tf.nn.sigmoid(tf.matmul(X, hidden_
                                         weights1)+ b)
  return tf.matmul(hidden_layer, ow)

dsX = np.linspace(-1, 1, trainsamples + testsamples).
                                            transpose()
dsY = 0.8*pow(dsX,2)+2*dsX+np.random.randn(*dsX.
                                  shape)*0.22+0.8

X = tf.placeholder("float")
Y = tf.placeholder("float")

# Create first hidden layer
hw1 = tf.Variable(tf.random_normal([1, 10], stddev=0.1))

# Create output connection
ow = tf.Variable(tf.random_normal([10, 1], stddev=0.0))

# Create bias
b = tf.Variable(tf.random_normal([10], stddev=0.1))
model_y = model(X, hw1, b, ow)

# Cost function
cost = tf.pow(model_y-Y, 2)/(2)

# construct an optimizer
train_op = tf.train.GradientDescentOptimizer(0.05).
                                     minimize(cost)

# Launch the graph in a session
epochs = 50

with tf.Session() as sess:
  tf.global_variables_initializer().run() # initialize all
                                            variables
```

```
for i in range(1,epochs):
  #randomize the samples to implement a better training
  dsX, dsY = shuffle (dsX.transpose(), dsY)
  trainX, trainY = dsX[0:trainsamples],
                                      dsY[0:trainsamples]

  for x1,y1 in zip (trainX, trainY):
    sess.run(train_op, feed_dict={X: [[x1]], Y: y1})

  testX, testY = dsX[trainsamples:trainsamples +
            testsamples], dsY[0:trainsamples:trainsamples+
                                            testsamples]

  cost1=0.
  for x1,y1 in zip (testX, testY):
    cost1 += sess.run(cost,feed_dict={X:[[x1]], Y:y1})/
                                          testsamples

    if (i%10 == 0):
      print "Average cost for epoch " + str (i) + ":" +
                                        str(cost1)

plt.figure()
plt.title('Original data')
plt.scatter(dsX,dsY)
plt.show()
```

Listing 5.3 starts with the same code as Listing 5.2, and the new code starts immediately after the definition of dsX and dsY. The new section defines the TensorFlow placeholders X and Y, followed by the TensorFlow variable hw1 (the first hidden layer) that is a one-element vector that in turn contains a 1x10 vector of random numbers. Next, the TensorFlow variable ow is the output connection, which is defined as a 10x1 column vector of random numbers. Then the TensorFlow variable b is initialized as a 1x10 fector of random numbers.

In case there's any doubt, the dimensions and the contents of hw1, ow, and b are shown below (your random values will be slightly different):

```
('hw1 shape:', TensorShape([Dimension(1), Dimension(10)]))
[[-0.00759747 -0.11399724 -0.03839068 -0.02203284
  0.00551973  0.04510954
  0.05798442  0.10647184  0.01218118  0.01847549]]
('ow shape:', TensorShape([Dimension(10), Dimension(1)]))
[[0.]
 [0.]
 [0.]
 [0.]
 [0.]
 [0.]
 [0.]
 [0.]
 [0.]
 [0.]]
('b shape:', TensorShape([Dimension(10)]))
[-0.16788599 -0.02116957 -0.08303189  0.1054974  -0.1958618
 -0.15026365
 -0.01868325 -0.2112294   0.07618438 -0.09956991]
```

The next portion of Listing 5.3 initializes the variable `model_y` by invoking the `model()` method with the previously defined variables `X`, `hw1`, `b`, and `ow`. The cost function is defined via the TensorFlow `pow()` method, and it's an MSE cost function. The `train_op` variable is initialized as the TensorFlow Gradient Descent Optimizer, with a learning rate of 0.05, and then an invocation of the `minimize()` method with the defined cost function.

After initializing epochs with the value of 50, the `with` code block initializes the previously defined variables, followed by a `for` loop that iterates 50 times. The `for` loop starts by initializing `dsX` and `dsY` with the result of "shuffling" `dsX` and `dsY`, which helps to reduce the chances of matching unintended patterns of data. The array variables `trainX` and `trainY` are assigned the first 200 values of the array variables `dsX` and `dsY`, respectively, where 200 is the value of the variable `trainsamples`.

The next code snippet in Listing 5.3 is a simple loop that trains the neural network in a pair-wise fashion via the numbers in the arrays `trainX` and `trainY`.

In a similar fashion, the array variables `testX` and `testY` are initialized with the rightmost 60 values of the arrays `dsX` and `dsY`, respectively, where 60 is the value of the variable `testsamples`.

Next, the `cost1` variable is initialized to zero, followed by another simple loop (similar to the previous inner loop) that calculates the value of the `cost1` based on the test-related data. This loop also prints the cost value after every 10 iterations.

The last portion of Listing 5.3 is the plot-related code that is the same code that you saw in Listing 5.2.

The output from Listing 5.3 is here:

```
Average cost for epoch 10:[[0.02131385]]
Average cost for epoch 20:[[7.2599596e-06]]
Average cost for epoch 30:[[0.0107516]]
Average cost for epoch 40:[[0.09003434]]
```

Notice the variability of the average cost in the preceding output.

The graph for the code in Listing 5.3 is almost identical to the graph from Listing 5.2 (except for differences due to small random values), so the graph will not be shown here.

WORKING WITH KERAS

As mentioned in the introduction, this section provides a *very* brief introduction to Keras that will give you the bare minimum for creating Keras-based models (indeed, entire books are devoted to Keras).

If you are new to Keras, there are a couple of points to keep in mind. First, Keras is integrated into the latest versions of TF 1.x as well as a core part of TF 2, located in the `tf.keras` namespace. Second, Keras is well-suited for defining models to solve a myriad of tasks, such as linear regression and logistic regression, as well as Deep Learning tasks involving CNNs, RNNs, and LSTMs.

A Keras model generally involves at least the following sequence of steps:

- Specify a dataset (if necessary, convert data to numeric data)
- Split the dataset into training data and test data (usually 80/20 split)
- Define the Keras model (`tf.keras.models.Sequential()`)
- Compile the Keras model (the `compile()` API)
- Train (fit) the Keras model (the `fit()` API)
- Make a prediction (the `prediction()` API)

Note that the preceding bullet items skip some steps that are part of a real Keras model, such as evaluating the Keras model on the test data, as well as dealing with issues such as overfitting.

The first bullet item states that you need a dataset, which can be as simple as a CSV file with 100 rows of data (and just 3 columns). In general, a dataset is substantially larger: it can be a file with 1,000,000 rows of data and 10,000 columns in each row. We'll look at a concrete dataset in a subsequent section.

Next, a Keras model is in the `tf.keras.models` namespace, and the simplest (and also very common) Keras model is `tf.keras.models.Sequential`. In general, a Keras model contains layers that are in the `tf.keras.layers` namespace, such as `tf.keras.Dense` (which means that two adjacent layers are fully connected).

The activation functions that are referenced in Keras layers are in the `tf.nn` namespace, such as the `tf.nn.ReLU` for the ReLU activation function.

Here's a code block of the Keras model that's described in the preceding paragraphs (which covers the first four bullet points):

```
import tensorflow as tf

model = tf.keras.models.Sequential([
  tf.keras.layers.Dense(512, activation=tf.nn.relu),
])
```

We have three more bullet items to discuss, starting with the compilation step. Keras provides a `compile()` API for this step, an example of which is here:

```
model.compile(optimizer='adam',
              loss='sparse_categorical_crossentropy',
              metrics=['accuracy'])
```

Next we need to specify a training step, and Keras provides the `fit()` API (as you can see, it's not called `train()`), an example of which is here:

```
model.fit(x_train, y_train, epochs=5)
```

The final step is the prediction, and Keras provides the `predict()` API, an example of which is here:

```
pred = model.predict(x)
```

Listing 5.4 displays the contents of `tf-basic-keras.py` that combines the code blocks in the preceding steps into a single code sample. Note that this code sample does not include the dataset, so you won't be able to launch the code (but you will in the subsequent section).

LISTING 5.4: tf-basic-keras.py

```
import tensorflow as tf

# NOTE: we don't have the train data and test data

model = tf.keras.models.Sequential([
  tf.keras.layers.Dense(1, activation=tf.nn.relu),
])

model.compile(optimizer='adam',
              loss='sparse_categorical_crossentropy',
              metrics=['accuracy'])

model.fit(x_train, y_train, epochs=5)
model.evaluate(x_test, y_test)
```

Listing 5.4 contains no new code, and we've essentially glossed over some of the terms such as the optimizer (an algorithm that is used in conjunction with a cost function), the loss (the type of loss function) and the metrics (how to evaluate the efficacy of a model).

The explanations for these details cannot be condensed into a few paragraphs (alas), but the good news is that you can find a plethora of detailed online blog posts that discuss these terms.

The next section contains a code sample that extends Listing 5.4 by providing the MNIST dataset for the Keras-based model.

WORKING WITH KERAS AND THE MNIST DATASET

Listing 5.5 displays the contents of `tf-simple-keras-mnist.py` that shows you how to create a simple Keras-based model for training the MNIST dataset.

LISTING 5.5: tf-simple-keras-mnist.py

```
import tensorflow as tf

mnist = tf.keras.datasets.mnist
(x_train, y_train),(x_test, y_test) = mnist.load_data()
```

```
x_train, x_test = x_train / 255.0, x_test / 255.0

model = tf.keras.models.Sequential([
  tf.keras.layers.Flatten(input_shape=(28, 28)),
  tf.keras.layers.Dense(512, activation=tf.nn.relu),
  tf.keras.layers.Dropout(0.2),
  tf.keras.layers.Dense(10, activation=tf.nn.softmax)
])
```

model.summary()

```
model.compile(optimizer='adam',
              loss='sparse_categorical_crossentropy',
              metrics=['accuracy'])

model.fit(x_train, y_train, epochs=5)
model.evaluate(x_test, y_test)
```

Listing 5.5 contains a new code snippet (shown in bold) that initializes the mnist variable from the MNIST dataset, and then populates x_train, y_train, x_test, and y_text with training data and test data, respectively. The first variable in each pair references the images, whereas the second variable in each pair references the corresponding labels for the images.

Notice that the values in x_train and y_train, are scaled down by a factor of 255, which means that the values are between 0 and 1 instead of 0 and 255 (the latter range is the allowable pixel values). Why is this done, you ask? Surprisingly, and perhaps non-intuitively as well, the model performs better when the input values are between 0 and 1 instead of 0 and 255.

Now launch the code in Listing 5.5 and you will see the following output:

Layer (type)	Output Shape	Param #
flatten (Flatten)	(None, 784)	0
dense (Dense)	(None, 512)	401920
dropout (Dropout)	(None, 512)	0
dense_1 (Dense)	(None, 10)	5130

```
Total params: 407,050
Trainable params: 407,050
Non-trainable params: 0
```

```
Epoch 1/5
60000/60000 [====================] - 14s 231us/step - loss:
0.2208 - acc: 0.9349
Epoch 2/5
60000/60000 [====================] - 14s 227us/step - loss:
0.0973 - acc: 0.9704
Epoch 3/5
60000/60000 [====================] - 13s 222us/step - loss:
0.0671 - acc: 0.9794
```

```
Epoch 4/5
60000/60000 [=====================] - 14s 227us/step - loss:
0.0538 - acc: 0.9821
Epoch 5/5
60000/60000 [=====================] - 13s 224us/step - loss:
0.0445 - acc: 0.9856
10000/10000 [=====================] - 1s 61us/step
```

If you look at the last line of the preceding output, you will see that the accuracy is 98.56%, which is quite impressive!

SUMMARY

This chapter started with an overview of activation functions, why they are important in neural networks, and also how they are used in neural networks. Then you saw a list of the TensorFlow APIs for various activation functions, followed by a description of some of their merits.

Next, you saw an overview of Logistic Regression, followed by a TensorFlow code sample involving Logistic Regression (and also uses the sigmoid activation function). Then you got a very condensed introduction to Keras, followed by an example of creating a Keras-based model that was trained on the MNIST dataset.

INDEX